D1520412

Exploring materials with young children

Exploring materials with young children

Roy Sparkes

B T Batsford Limited   London and Sydney

© Roy Sparkes 1975
First published 1975
ISBN 0 7134 2926 7

Designed by Charlotte Gerlings
Libra Studios
Filmset in Monophoto Garamond (156)
by Keyspools Ltd, Golborne
Lancashire
Printed in Great Britain by
William Clowes and Sons Limited
Beccles, Suffolk
for the publishers
B T Batsford Limited
4 Fitzhardinge Street
London W1H 0AH and
23 Cross Street, Brookvale
NSW 2100, Australia

Photographic work by
Derek Clements

372.5
Sp 26e

Contents

Acknowledgment

I am grateful to the teachers, children and friends who knowingly or unknowingly have contributed to the making of this book. My thanks are due to Mrs R T Francis, Headmistress, and the staff and children of Montgomerie Infants School, New Thundersley, Benfleet, Essex; to Mrs Wishearn, Mr Baldwin, the staff and children of St Pauls Wood Primary School, Orpington, Kent; to John M Pickering for figure 10; Guy Scott and Tony Dockerill for figure 44 and to Thelma M Nye and Harriet Murray-Browne for their encouragement and advice.

This book is dedicated to Neil and Claire

Introduction

Here is a source book of ideas and techniques for use with young children in their play/learning activity. It is concerned mainly with the use of materials that are to be found ready to hand in the home and at school. Often these are waste and discarded objects.

It is considered here that although an end product, such as a construction, can be important, the activity and the thinking that goes into producing it is the more valuable. With young children an activity may be a spontaneous reaction in a particular situation. Information and suggestions are given throughout this book in order that parents and teachers may have a knowledge of processes and a supply of materials available.

It is hoped that the ideas given here will be used as starting points to help parents and teachers to stimulate imaginative thinking in play/work, and encourage children to make their own discoveries.

The materials have been listed alphabetically. Cross references have been made so that a description of each process occurs once. Further information on techniques can be found in the Bibliography.

Technical notes

COLLAGE/APPLIQUÉ

There is often confusion in the use of these two terms. They are both concerned with the application of materials to a surface which is usually flat and two-dimensional. Throughout this book *appliqué* has been reserved to describe the use of fabrics, either sewn or stuck to a fabric base, while *collage* is employed to describe the use of any other combination of materials.

RELIEF

Whereas with collage the two-dimensional nature of the base surface is preserved, in making a relief panel the materials fixed to the base may project away from it to a range of different heights. A strong base is needed.

BASEBOARD

This is used especially for relief panels and collages. Where heavier or more pliant or fragile materials are being used, or when building away from the flat surface, as in a relief panel, a strong base is necessary. Hardboard will serve well for most work and offcuts may be easily found. Chipboard, blockboard, plywood, and Essexboard are less readily available as offcuts and tend to be heavier. They all give a more stable base than hardboard, although wooden battens can be nailed, screwed, or glued to the back of hardboard to strengthen it. Strawboard can be used in lighter and small scale work.

Collage designs can be made on strong paper, card, strawboard or the surfaces of most waste cartons.

PAPERS

There are many different kinds available, some of which are expensive. For most drawing, painting and printmaking processes, newsprint or kitchen paper will serve well. In many cases waste papers and cards can be used: the backs of envelopes, wrapping paper, insides of cardboard boxes, and the reverse side of most wallpaper offcuts. It would be worthwhile, whenever the opportunity occurs, to use a better quality paper, such as a medium grade cartridge paper, as this usually produces a better result. Sugar paper is versatile and strong, and can be obtained in many different colours. Tissue papers, coloured gum-backed papers, and different grades of card are useful for certain activities, and offer the experience of different textural and colour qualities. All kinds of papers, whether previously used or not, are of value in imaginative work and particularly in collage designs.

1 Coloured, gummed paper, fixed to a sugar paper base, has been used to create a tortoise. The rubbing was taken from the collage of the tortoise, using a thick wax crayon and kitchen paper (newsprint)

2 *Paper offcuts have been used by a six-year-old boy to make a linear design, a collage stuck onto a sugar paper base*

3 *The clown is a favourite subject of young children. This one was made from different papers and cards pasted to a sugar paper base. The nose is a segment from an egg box*

4 *This 'giraffe', the work of a group of five-year-olds, is a collage picture consisting of rubbings from different surfaces in a room. Each child collected rubbings and contributed them to the collage. Kitchen paper and thick wax crayons were used and the adhesive was* Polycell

PRINTMAKING

In the early stages printmaking will mainly be concerned with discovering what different kinds of marks can be made. Various factors can influence this: the textural qualities of different surfaces; the ink, dye, or paint used; the method of printing; and the paper or other surface on which the print is to be made.

At first the prints will be the result of play and discovery. The marks may be somewhat haphazard. Gradually children can be encouraged to arrange the marks into a design or pattern.

SORTING AND DISCUSSION

These are of value in developing a critical awareness and the ability to categorise, select and differentiate, as well as facilitating an ordered, efficient way of storing and working.

In discussions children can share experiences and ideas which may feed into their play activity, and help to develop an awareness of the true potential of the materials they use.

Waste boxes are useful for the storage of collected and sorted items. Shoeboxes, matchboxes, biscuit tins, and chocolate boxes are particularly useful.

ADHESIVES

Cold water pastes, under various trade names, are satisfactory for general purposes and can be kept for some time if mixed with a few crystals of permanganate of potash. *Gloy*, *Uhu*, and *Copydex* can be used for most papers; and *Copydex* and *Caretex* for most textiles. *Cow Gum* is useful where it is necessary to pull apart surfaces previously joined. *Evostick* is a strong impact adhesive; it reacts against polystyrene. *Unibond* however can be used with polystyrene. *Gluak* holds immediately, and is useful when making lightweight constructions as are *Uhu* and *Caretex*. PVA binders, such as *Marvin Medium* and *Evostick Woodworking Adhesive*, are good all-purpose adhesives which can be used with polystyrene and are particularly useful for collage work. They can be diluted when necessary. *Polycell* is a good seal for most porous surfaces, particularly for thinner papers; it can be used in papier mâché work, and it keeps well. There are many different adhesives available. It is important to read directions carefully before allowing children to use them.

CONSTRUCTION

This term is used when fixing different pieces of one or more materials to each other to build a three-dimensional form, which may be free standing or fixed to a base. Fixing may be by gluing, clipping, pinning, slotting, or interweaving.

Constructions can be made from a wide variety of materials, ranging from paper to metal.

BRICKS AND STRUCTURES

It should be possible to supply all the bricks children may need for building games from waste cartons and offcuts. It is important to use a limited number of sizes or units, since young children may find it difficult and frustrating to have too wide a range of different proportions. Cardboard boxes (hollow) are good for large structures. Offcuts of non-splintering wood, such as beech, make good smaller blocks for building games; rough edges can be smoothed with sand or glass paper; they can be coloured with emulsion paint or with powder paint mixed with a PVA binder. Strips of wood can be useful for roads, bridges, roofs and floors.

Discussions could centre on the use of mathematical terms: taller than, too long, shorter than, too short, too high, twice as long, larger than, smaller than, heavier than, lighter than. Children may wish to talk about their fantasies and imaginative experiences which occur in the process of play.

SAND

Outdoor: a ventilated cover, for example wire netting, is necessary to keep cats out when not in use. It should have adequate drainage. Dirty sand can be washed with a hose.

Indoor: silver sand is best as it is fine and non-staining.

A child with a skin complaint or some form of eczema should have medical approval for playing with sand.

Points for discussion might be: what does the sand feel like – damp, dry, rough, gritty, smooth? Damp sand is heavier than dry – why? What happens when damp sand becomes dry? Patterns can be made in damp sand. Dry sand can be sifted; it can be poured from one container to another.

WATER

Try using pure water, and water with additives such as washing-up liquid, or blue bag in order that the children may experience the changes in colour, consistency and texture. Water for children to play with should be warm and placed in a warm area. Aprons should be worn.

Discussions might begin with the concepts of full and empty; floating and sinking; absorbent and non-absorbent; melting, dissolving, and solidifying; rusting and non-rusting. Examples of melting can be seen in ice, ice-cream and ice-lollies; an example of dissolving, water to steam; and of solidifying, making a jelly.

PROTECTIVE CLOTHING

Children should be encouraged to wear aprons or overalls. Clothing may need to be protected from paints, pastes and glues, and water splashes. Plastic or rubberised aprons are suitable. A clean 'cast-off' will serve well – an old shirt or pyjama jacket.

DISPLAY

Display areas provide source material for learning. Children will want to go to a place/room to play and to make and look at things if it contains interesting, well-presented displays which are changed from time to time. 'Colour corners', nature study tables, displays of objects of a similar material such as wood or wool, a 'touch table' or a 'smell table', objects collected from a particular place such as the beach, a water tank with collections from pond life – all are starting points for learning and may stimulate interests and ideas for the children.

With practice what is good display will become increasingly evident. A mount leaving a 25 mm (1 in.) border all round would look well on individual pictures. Space between pictures and designs is important. This need not be more than 50 mm (2 in.) although some work will look better in a larger space. It is usually best to arrange work so that colour harmonises from one picture to the next. Contrasting colours can be useful for a particular effect, but there will be a tendency for them to jump or vibrate, especially if complementary colours – red and green, blue and orange, yellow and purple – are placed in juxtaposition. Neutral backgrounds – white, grey, or black – will help colour to be seen to good advantage.

Interesting and eye-catching displays can be made when work is of different sizes, as long as the horizontal edges are parallel to each other and there is an eye line, for example the tops of pictures.

There are various means of fixing two-dimensional work to a display board. The best are staple guns or 'tackers', and pin-pushers. Staple guns which fire into a flat surface are ideal. It is safer not to allow young children to use a staple gun. Staple extractors can be purchased, but a pair of pliers is as easy to use if the staples are fired into the mounting board at an angle, leaving a corner projecting slightly.

Pin pushers are useful for displays of lightweight work. Dressmaking pins are placed into the barrel and then pushed into the display board. Pins and staples are preferable to drawing pins, which fall out easily and tend to be unsightly, obtruding on a display.

In the home, a pinboard on a bedroom wall and a shelf for three-dimensional objects and books would help a child to feel that what he does in his play/work activity has a place and is welcome.

Cleanliness and tidiness are important and will help children to organise their thinking, and will facilitate their learning.

Balloons

Use those left over from parties, as the cost of buying them may be prohibitive when working with a class of children.

PAINTING/DRAWING

Draw or paint direct onto the balloon surface. Try faces, human and animal. Felt-tip pens (non-toxic) are good for this.

COLLAGE

Use fabric scraps and coloured papers to decorate the surface of balloons, gluing with *Caretex* or *Cow Gum*.

PAPIER MÀCHÉ

Blow up the balloon and tie. Tear old newspapers into strips (newsprint and tissue paper could be used) and soak in a bowl of cold water paste or *Polycell* until the paper is saturated. Cover the balloon in five or six layers. In the application of each layer let the strips overlap. Each layer must be carefully stuck down. To give a stronger structure strips of fabric could be used for one or two layers (eg a fine muslin). When the papier màché is dry, the balloon can be

pricked and removed through the hole at the top. The hole can then be covered with strips of paper màché. The structure can be decorated, using a collage technique (tissue papers are good) or painting and drawing. Holes can be cut out if desired and these may be enjoyed simply for their shapes, or be further fashioned to suggest eyes, nose and mouth. An alternative would be to form physical features using papier màché to build out from the basic structure.

To make a mask cut the basic papier màché structure into halves. Two children working together can then each have one half. Mouth, eyes and nose spaces can be cut. A bridge can be built across the hole for the nose, using a card foundation and more papier màché. The mask can be decorated, using collage, painting and drawing. Elastic can be fitted to go round the head. Or, a stick (cane or dowelling rod) with which to hold the mask can be fixed with *Sellotape* (*Scotch Tape*), brown gum strip and/or an adhesive (an impact adhesive such as *Evostick*). Papier màché can also be used to fix the stick.

5   *Pasting strips of newspaper over an inflated balloon. An old shirt makes a satisfactory overall*

Ball-point pens

In addition to their use as drawing/writing instruments, ball-point pens can be used in a number of ways.

PRINTMAKING
Take prints from the end of the outer casing which will make small rounded marks. Use a sponge inking pad. See PRINTMAKING under SPONGE.

BLOW PAINTING
The outer casings make good tubes with which to blow ink and paint blots across a sheet of paper. Ensure the tubes are clean before use.

CONSTRUCTIONS
Make a free standing form with the outer barrels and/or the discarded centres. Fix one at a time, using an impact adhesive.

RELIEF
Arrange the used pens into a design and glue to a base board. They can be interspersed, placed close together, and built up in layers across each other or one on top of another.

6 *'Blow painting'. Claire, aged three, is using a ball-point pen barrel to blow ink across a paper surface. A water colour brush was used to make ink drops on the paper. Old newspapers protect the table surface*

Bark

Collect different kinds of bark from the countryside. It is important to train children to take only that which is on the ground or on fallen trees.

DISCUSSION
1 *Colours* Some sorting can be done. See SORTING under BEADS. The value here would be in observing variations in colour and in naming colours, as well as the kinds of trees from which the bark came.
2 *Sense of touch* What does the bark feel like? Does the texture feel the same as it looks?

RUBBINGS
See RUBBINGS under WAX.

7 Bark and driftwood may be enjoyed for its shape as much as for its texture and colour

RELIEF
Choose pieces of bark that will lie reasonably flat. The colours, shapes and textures will affect the finished appearance. Arrange them into a design and glue to a firm base, *eg* hardboard or strawboard.

PRINTMAKING
Select pieces of bark that have interesting textures and that are reasonably flat. Cover the bark with water based printing ink or powder colour using a brush, or use a sponge inking pad. See PRINTMAKING under SPONGE. Either place a sheet of newsprint over the bark and press with the hand, or, holding the bark in the hand, press it, colour downward, onto the paper. The backs of large envelopes, fabric scraps, and opened-out paper bags can be used as surfaces on which to print.

Beads

Collect a wide variety of beads.

SORTING
Boxes and empty waste cartons, *eg* shoe boxes, will be needed to house the different kinds of beads.
1 *Colour* Put all the reds together, all the blues together, etc using one box for each colour. Further investigation and sorting could be within one colour area, *eg* different kinds of reds.

Beads from one colour box could be chosen to make a tonal line, placing the darkest red at one end and the lightest red at the other end.
2 *Scale* Choose 'the biggest' and 'the smallest' at first. Once a child has grasped this concept, investigation of more subtle differences could be carried out. A line of beads could be made with the largest at one end and the smallest at the other. Questions related to the sizes of the beads in the line might be asked.
3 *Shape* Sort the beads into sets of basic shapes: circular, square, oval, triangular.
4 *Materials* Sort into categories: glass, plastic, wood, bone, stone.

MOSAIC

Use lids of waste cartons (*eg* cheese boxes) and *Plasticine* or *Newclay*. The *Plasticine* can be pressed into a lid mould to make a base. The beads can then be inlaid by pressing with the fingers. The beads can be touching or spaced apart to make a design. The nearer they are together the firmer the surface will be. If *Newclay* is used, it would be advisable to paint the finished mosaic with a clear varnish or a PVA binder, *eg Marvin Medium*. This will obviate the tendency for beads to fall out as the moisture in the *Newclay* evaporates.

APPLIQUÉ

Use the beads in combination with fabric scraps applied to a paper, card or fabric base. Children will find all manner of imaginative uses for them, *eg* eyes, a nose, bobbles on hats, wheels, and buttons on jackets. They can also be used to decorate three-dimensional work.

LOW RELIEF

Arrange the beads into a design and then glue to a board or strong card. Some of the larger beads might be used to contrast with the low height of the smaller ones.

8 *Sorting beads according to size and placing them in used plastic containers*

9 *Detail of a design based on a spiral motif, beads and buttons were pressed into* Plasticine *by a five-year-old girl. The* Plasticine *was arranged on thin card which was then mounted onto sugar paper. Further spirals were drawn with a felt-tip pen*

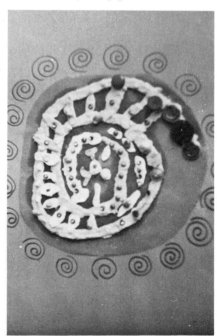

Bones

Use bones that have been cleaned, *eg* by boiling.

DISCUSSION
Some starting points are:
What does the shape look like?
Suggestions from children have included a gun, a hammer, and a fish.
Which animal does it come from?
How and where does that animal live?
What does the bone feel like?
How does it make you feel when you look at it? Happy, sad, afraid, cold, like jelly, have been suggested by children.

RELIEF
Choose bones that look well together to make a design. Arrange and glue them to a firm board base, *eg* a hardboard offcut.

10 *A bone found on a walk along a beach is the starting point for discussion*

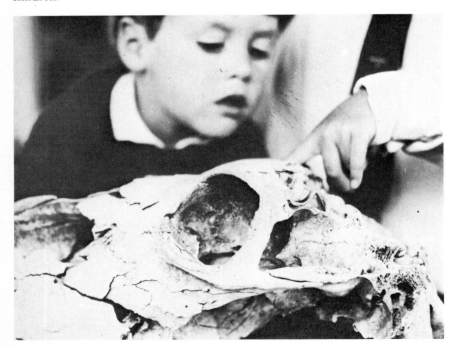

Bottle caps

Collect many different kinds of sizes, colours, and materials.

SORTING
Children can sort into categories.
See SORTING under BEADS.

PRINTMAKING
1 Use a sponge (foam plastic) pad, and water based paint or ink in a shallow dish or palette. See PRINTMAKING under SPONGE. Both tops and bottoms of the caps can be tried, printing on paper, card (in sheet form), waste cardboard articles (*eg* sides of boxes and the inside of box lids), foil, and the backs of envelopes. The bottle cap, laden with colour, should be pressed on to the surface to be printed.
2 See IMPRESSIONS, below.

RELIEF
Using metal and/or plastic caps, arrange them into a design and then glue to a board or card base. They could be used either way up, and be placed at various distances from each other. Areas of board not 'capped' could be painted or left in its natural state. There could be two or three layers, built up in certain areas.

COLLAGE
Shallower caps and those of a more pliant material such as milk bottle tops (which will need cleaning)

could be used in a design which preserves the two-dimensional quality of the base to which they are fixed. A strong paper or card would suffice as the base.

The imaginations of young children can employ bottle caps in all kinds of ways in two- and three-dimensional work, for example as eyes, hats, wheels on the sides of boxes for cars and barrows, and as coins.

MOSAIC

Make a pattern with the caps, pressing them into *Plasticine* or *Newclay*. See MOSAICS under BEADS.

IMPRESSIONS

1   Give the surface of *Newclay*, *Plasticine* or *Play dough* a rich texture by pressing bottle caps into it, then removing them, leaving their impressions. A block or slab can be made and a pattern impressed. The impressions can be interspaced and overlapped.

2   Print from the impressions made in a small block. See PRINTMAKING under BARK. The handheld block, its face covered with colour, is pressed on to the surface to be printed. Using smaller pieces of *Plasticine*, insignia (stamps) can be made.

11   *Caps from various bottles have been glued with* Marvin Medium *to a blockboard offcut by a seven-year-old. The reflective qualities of the caps and their different sizes add interest to the family of shapes*

Bottles

Collect both glass and plastic.

STORAGE
Particular items for which they are useful are: inks, paints, dye, paste and glue mixed ready for use.

12 *Ashley chooses from a display of bottles of different materials, shapes and sizes which invites children to use and develop their sense of smell. Discussions of children's experiences and preferences can be worthwhile. At the same time, holding and examining the different bottles provides an opportunity to develop the concept of form*

13 *Strips of newspaper have been stuck onto a bottle with a cold-water paste. A bottle with an even surface and straight neck can be used as a base form for papier mâché figures. Five or six layers of papier mâché are necessary for a strong form*

18

ROLLING PIN

Use them for making slabs of clay, *Newclay* and *Play dough*. Straight sided bottles are suitable. Care must be taken with glass bottles which may best be used only by parent and teacher.

CONSTRUCTIONS

Plastic bottles only should be used. There is an immense range of possibilities, for example, towers, tank gun turrets and binoculars. See CONSTRUCTIONS under PLASTIC CONTAINERS.

PAPIER MÂCHÉ

A bottle with an even surface and straight neck can be used as a base form for papier mâché people or creatures. Five or six layers of papier mâché are necessary for a strong form, and should be pasted on with *Polycell* or a PVA binder thinned with water. For the head, screw newspaper strips into a ball and bind with strips and paste. When the body is dry, pull out the bottle and fix the head to the body with further newspaper strips and paste which form the neck. When a satisfactory form has been built and dried it can be painted or decorated with fabric scraps and other waste items.

ROLLING PIN

Cut a 400 mm (15½ in. approx) length from a discarded broom handle. Smooth the ends with sandpaper. Use it to roll out clay, for example when making ceramic tiles and 'slab-pots'. One broom handle will make four to five rolling pins.

HOBBY-HORSE

Attach the horse's head to one end of the broom handle using string, glue and/or staples (the latter to be used by the teacher or parent). The horse's head can be made by padding out an old, clean sock or a bag with paper, fabric scraps or sponge. Fabric scraps can be glued or sewn to the outside for the facial details; and strips of fabric could be glued along the top of the handle as the horse's mane. Coloured rope or ribbon can also be glued to the handle for the reins.

14 *Neil, aged six, riding his hobby horse. The basis for this was a sock and a broom handle. The sock was stuffed with fabric scraps to give it the desired form for the horse's head. It was fixed to the broom handle by binding it with insulating tape (a fabric covering to conceal the tape may be desired). The mane is of wool and the eye and mouth are made from fabric scraps*

COLLAGE

Use either buttons only or buttons in combination with other materials. Make a design to be stuck onto a two-dimensional surface or, to decorate constructions and models, *eg* eyes for puppets (stuck or sewn – take care with needles).

PRINTMAKING

Fix a button with an interesting texture to a block so that it is easy to hold (glue to a wood block, or press onto a small block of *Plasticine*). Print, using a sponge pad. See

PRINTMAKING under SPONGE. 'Toggle' buttons can be handheld without fixing to a block. Press the button laden with colour onto the surface to be printed. Try making a repeat pattern.

SORTING

See under BEADS. Sort into sets according to colour, size, shape and texture.

PATTERN/DESIGN PLAY

Use the buttons to make patterns, leaving them unstuck so that changes and different patterns can be made. Work on a sheet of paper to give a framework to the patterns. When a child is satisfied with a design, it could be fixed into position with a PVA binder or *Evostick*.

MOSAICS

See MOSAICS under BEADS. Broken and whole buttons could be used.

IMPRESSIONS

See IMPRESSIONS under BOTTLE CAPS.

15   *Detail of* Button flowers, *by a seven-year-old girl.* Marvin Medium *was used to fix the buttons to the base (blockboard) and to each other. The girl chose the buttons herself from a collection in a box and showed concern for their size and contrasting colours*

Cane

PUPPETRY
Use a short length of cane as a rod for a puppet. See PUPPETS under COTTON REELS and TWIGS.

MASKS
Use to make a 'stick mask'.
1 See PAPIER MÂCHÉ MASKS under BALLOONS.
2 Cut a full face shape or an eye mask out of card. Seven and eight year old children should be able to cope with this; younger children will be happy simply painting/drawing a face on a rectangle of card. Spaces can be cut or torn out of the card for eyes, mouth and nose by the younger children, or for them if necessary. Paint/draw details and patterns on the mask shape. Some collage work could be incorporated. Attach a length of cane to the back of the card so that it protrudes at the bottom and can be held in the hand. It can be attached to the card with glue, *Sellotape* or brown gum strip. The mask could be of a human, animal or make-belief character.

RELIEF
Arrange cane of the same or different lengths and thicknesses into a design, and glue to a base board.

CONSTRUCTIONS
Make a free standing form, fixing the lengths of cane together one at a time, using an impact adhesive. Structures could be made without an adhesive by overlapping and interweaving.

COLOUR BOX
See COLOUR BOX under CARDBOARD BOXES.

DRAWING STICK
See DRAWING under SPONGE.

16 *A length of cane has been used by a seven-year-old child to make a stick mask. The cane was glued to the back of the card. Sellotape held the stick in place while the adhesive dried. The decorations include milk bottle tops, a bead, feathers, a cone and fabric scrap fixed with a PVA binder*

Cardboard boxes

Collect all shapes and sizes from matchboxes to those of trunk size.

IMAGINATIVE PLAY
1   To sit in as, for example, cars, buses, boats and space ships.
2   To wear as, for example, armour or robot suits.

The boxes can be decorated by painting and drawing, or with a wide range of waste materials glued to the surface. Cut holes for arms, legs, eyes and head as needed.

LIGHT EXPERIMENTS
1   A child in a large box, sitting on a small stall or kneeling, can look through a curtain hung over the opening. The size of the mesh of the fabric will determine how much light can penetrate and will affect the tonal qualities in the room as seen by the child. The denser the mesh, the darker the tonal quality. Old net curtains, stockinette, organdie, butter muslin, cotton sheeting, hessian would give different sizes of mesh.
2   *Lanterns*   Use long boxes *eg* certain cosmetics and sweet boxes. The cardboard holders for canned beer are particularly good since they are slotted together and can be easily reversed and re-slotted. Cut away the top and bottom – with canned beer holders the ends are already open. Cut windows or shapes and cover with coloured tissue paper or cellophane. It is better to open out the boxes so that the coloured transparencies can be fixed to the inside with glue or *Sellotape*.
a   Stand a night-light or candle in a jar that is taller than it in a safe place *eg* on a window sill. Place the lantern over the jar with an open end at the top.
b   The lantern box could be used as a mobile. A short rod or stick can be placed across the top of the lantern through small holes 25 mm (1 in.) from the top to join opposite sides. A

17   *This train is constructed entirely of cardboard waste – boxes and toilet roll centres. A pipe cleaner links the two parts together. The train could be painted and additional decorations made. The six-year-old who made this train decided he wanted to keep it undecorated*

thread is tied round the centre of the rod to suspend the lantern box.

TOUCH BOX

Place an object inside a box that is completely closed except for a hole, cut in one side, large enough to put both hands through. A child can feel the object and describe what is felt. Seven and eight year olds may like to draw or paint, or make a construction from the touch experience. The object could be shown to the children once they have described it in words or in practical work.

SCULPTURE WITH STRING

Remove the lid or one side of a box. Stretch string (various threads can be used including wool, twine, cottons and raffia) across the opening. Different directions can be taken to make interesting patterns. The natural colour of the string can be preserved or it can be painted or dyed. To fix, either knot both ends of each strand through holes made in the sides of the box; or weave across the opening through the holes making only two knots; or take the string all the way round the box tying each time or only once; or use *Sellotape*.

CONSTRUCTIONS

All manner of waste boxes will be useful and may help to stimulate imaginative play. Examples seen are

castles, houses, machines, cars, tanks, robots, armoured cars, prams and boats. Impact adhesives and PVA binders are good for gluing. Cold water paste is cheaper, but less durable; it is suitable for construction work combining boxes with papier mâché using newspaper strips. See PAPIER MÀCHÉ under BALLOONS. Collect boxes of one kind and size. Use them to make a unit structure.

18 *Cardboard boxes with spaces for windows and doors are used by six-year-olds to make animal houses for a model zoo. The base can be a sheet of hardboard or plywood*

23

RELIEF BASES

1   Lids of boxes can be used as bases for relief work. Collections of objects and waste materials can be arranged into a design within the framework of the lid and glued to it.

2   Lids can be used as bases for mosaics, see MOSAIC under BEADS.

COLOUR BOX

Experiment with complementary colours. Paint the interior of a box neutral – grey or white. Stretch red string across the opening of the box, or use lengths of cane or dowelling rod. The string or rods should be set back from the opening. They could be at different distances from it. Over the opening a sheet of green cellophane is then fixed. Note the effect of the cellophane on the string or rods. The other complementary pairs could be tried: blue and orange, yellow and purple.

PICTUREMAKING

1   Paint the outside of boxes. Different sides could be used to tell the parts of a progressive story or to make a series of patterns or colours. Or paint faces, animals, places, etc.

2   Flat surfaces cut or torn can be used separately for

a   painting and drawing

b   collage and relief work

3   Use strips of card as paint and paste applicators. Interesting effects can be produced by using a card applicator instead of a brush. Try a strip of card with a serrated edge.

PRINTMAKING

Boxes can be made into 'waste-blocks' for printing. Waste materials chosen for their textural qualities can be arranged into a design and then stuck onto a cardboard base. There should be very little variation in the height of the materials. Paint the surface with a water based colour or roll up with a water based printing ink. Squeeze a little ink from the tube onto the tile – formica offcut, lino, or other non-absorbent material – approximately 200 mm (7¾ in.) square. Roll the ink across the tile until it is of an even consistency on both the roller and the tile. Transfer the ink on the roller to the surface of the waste-block (or object to be printed) by rolling across it. When the roller is not in use put it back on the tile. Place a sheet of newsprint over the surface to be printed. With hand pressure (try with a soft cloth in the hand) on the back, take a print.

STORAGE

There are many ways in which boxes can be used for storage, especially for scrap and collected waste materials. They should be clearly labelled in bold lettering.

Both kinds have many uses: the peg which has two prongs and is 'clipped' over fabrics, and the peg which has to be pressed to open the end which 'clasps' the clothes.

TIE AND DYE

Pegs can be used to mask areas of the fabric (cotton sheeting is suitable) from the dye. The 'clasping' peg is more effective. Do not leave the fabric in the dye for more than a few seconds or the dye will be absorbed by the fabric under the pegs. The fabric could be folded or pleated before placing pegs in position. Try different folded patterns, see opposite.

19   *Clothes pegs have been clasped over a piece of white cloth which Neil, aged six, had previously folded. The pegs will mask areas of the cotton when it is placed in the dye. An old saucepan or bowl are suitable to hold the dye solution; if possible they should be enamel surfaced*

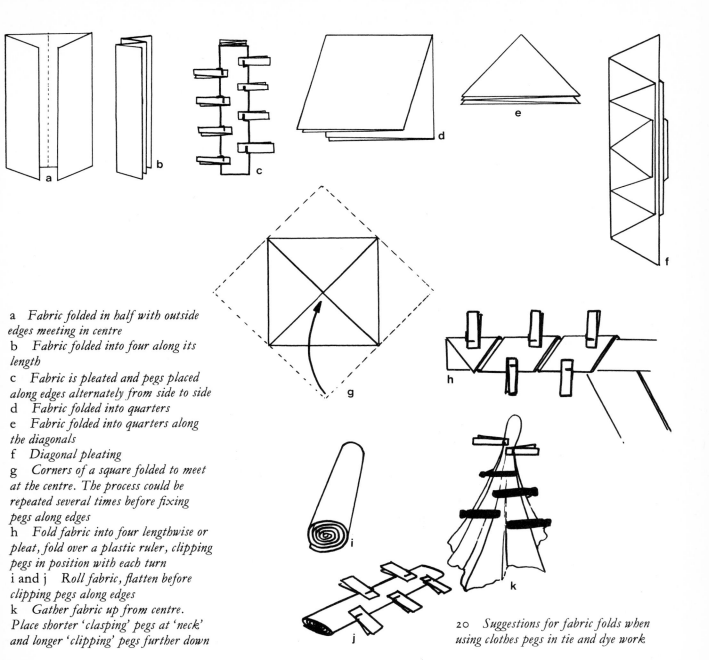

a   Fabric folded in half with outside edges meeting in centre

b   Fabric folded into four along its length

c   Fabric is pleated and pegs placed along edges alternately from side to side

d   Fabric folded into quarters

e   Fabric folded into quarters along the diagonals

f   Diagonal pleating

g   Corners of a square folded to meet at the centre. The process could be repeated several times before fixing pegs along edges

h   Fold fabric into four lengthwise or pleat, fold over a plastic ruler, clipping pegs in position with each turn

i and j   Roll fabric, flatten before clipping pegs along edges

k   Gather fabric up from centre. Place shorter 'clasping' pegs at 'neck' and longer 'clipping' pegs further down

20   Suggestions for fabric folds when using clothes pegs in tie and dye work

CONSTRUCTIONS
Make a three-dimensional free
standing form, fixing the pegs
together one at a time with an impact
adhesive. The longer, 'clipping' pegs
may also be fixed together without an
adhesive by slotting them one into
another.

Individual pegs can be drawn on
and painted to make 'peg people'.
Felt tip pens are good for drawing
facial details. Pieces of fabric can be
glued on as clothes, hair, etc.

RELIEF
Arrange the pegs into a design and
glue to a base board.

DRYING LINE
Hang up paintings and prints to dry
on a line stretched across a part of the
room.

DRAWING BOARD CLIP
Use pegs to clip paper to a board
whilst drawing. The smooth side of a
piece of hardboard will make a good
drawing board thin enough for pegs
to be clipped over easily.

PRINTMAKING
Print from the end or the length of a
peg using a sponge inking pad. See
PRINTMAKING under SPONGE. Try
making a repeat pattern with both
kinds of marks.

CLAY MODELLING TOOLS
Break apart the two prongs of the
'clipping' type of peg. Smooth any
rough edges with sandpaper. This will
provide two modelling tools for
work with clay and other modelling
materials.

Pegs used whole will make
interesting marks for surface
decoration on clay work.

21   *The seven-year-old boy who made
this 'horse' has clipped and slotted the
clothes pegs together. No adhesive has
been used*

26

Cocktail sticks

Both the wood and plastic kinds, plain or coloured, can be used.

SGRAFFITO
See SGRAFFITO under WAX. Cocktail sticks are useful for scratching, carving and making impressions in surfaces.

RELIEF
Arrange the cocktail sticks into a design and glue to a base card or board.

CONSTRUCTIONS
1   Make a free standing form, fixing the sticks together one at a time. *Caretex, Uhu* and *Evostick* are good adhesives for this work.
2   Use the sticks to decorate the surfaces of constructions, such as houses, boats and stables, made in other materials.
3   Use them to join units together to make a unit structure. See CONSTRUCTIONS 3 under EGG

BOXES. A unit might be an egg box segment or a cork.
4   Use them as arms and legs. They can be pressed into cork, potatoes and other materials which are being used as bodies.

22   *Using a cocktail stick in sgraffito work*

Cork

Collect different kinds, for example floor tiles, bottle stoppers, table mats, and cork floats used in fishing.

SORTING
See SORTING under BEADS.

DISCUSSION
Where do we get cork from? What is it used for? Study the patterns and textures, the children describing what they see and touch.

CONSTRUCTIONS
1 *Figures* Use matchsticks or cocktail sticks, which can be easily pushed into the cork, for arms and legs and to join pieces together for head and body.
2 *Structures* *a* For imaginative play *eg* as supports for bridges; as towers (corks built one on top of another); and as funnels for ships.

b For unit structures, using corks (bottle stoppers) of similar dimension and colour.

RELIEF
Arrange into a design and glue onto a base board. Various thicknesses and textures of cork could be used. Offcuts of cork floor coverings give good results. Bottle stoppers can be cut into sections (by parent or teacher). An impact adhesive or a PVA binder is the most suitable.

PRINTMAKING
1 Hand-held corks (stoppers) can be used to make a printed repeat pattern. A sponge pad is suitable for this. See PRINTMAKING under SPONGE.
2 Thinner sheets or pieces of cork would add interest to the textures in a 'waste-block' print. See PRINTMAKING under CARDBOARD BOXES.

RUBBINGS
See RUBBINGS under WAX. Make rubbings from
a natural cork
b collages or relief panels

23 *Cork bottle stoppers have been joined together using matchsticks to make this man. The six-year-old boy who made this drew the face with a felt tip pen. Cocktail sticks are more easily inserted into the cork than matches*

Corrugated cardboard

RUBBINGS
Lay a sheet of clean newsprint over the fluted surface and rub over it carefully and firmly with a thick, coloured wax crayon. See also under COLLAGE (below).

COLLAGE
1   Tear or cut into strips and shapes. Place them into a design on a paper or card base with the flutings in different directions. Take a rubbing.
2   A combination of corrugated cardboard and other materials can be made into a collage.

PRINTMAKING
1   Print from the collages. See PRINTMAKING under CARDBOARD BOXES.
2   Print from tight rolls (corrugations can be outside or inside the roll) tied with string or elastic bands. Prints can be taken from the end section and from the side rolled across the paper. Use a sponge pad. See PRINTMAKING under SPONGE.

DISPLAY
Use sheets of corrugated cardboard as background or ground base for displays of children's work.

CONSTRUCTIONS
1   Make cylinders, fixing the overlap with glue, paper clips, and/or *Sellotape* or brown gum strip. The cylinders may become imaginary lighthouses, towers, hats, or whatever children require to play out imaginative thinking.
2   Use as sheet material either in two dimensional form *eg* as rafts, matting, and ploughed fields; or in standing structures, glued, clipped, or slotted together for houses, sheds, castles, shops, garages, etc. There are many possibilities. Constructions made simply for the sake of exploring the materials are also worthwhile. The children will be learning about the properties of the materials, points of balance, and fixing techniques; and gaining manual dexterity in processes such as cutting, bending and folding cardboard.

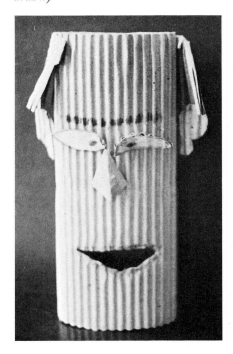

24   *Corrugated cardboard made into a cylinder with smaller pieces glued or* Sellotaped *to the surface and a shape cut out for the mouth (it could be drawn)*

Cotton reels

Collect both wood and plastic reels.

COLOUR COLUMNS
Paint each reel with powder colour (preferably mixed with a PVA binder). Some of the reels could have patterns painted on them. Build the reels one on top of another, and in different arrangements such as a pyramidal structure. Single columns

25 Fabric scraps have been glued to two cotton reels by a seven-year-old to make this creature. The cotton reels could be fixed together

could be made in reels of one colour to give practice in colour categorising. Children might try placing the reels in a certain sequence. For practice in number work, ask the children to put the reels into sets, for example five reels of each colour.

A more permanent arrangement of the coloured reels can be made by inserting a length of dowelling rod or bamboo cane through the holes. *Sellotape*, insulating tape, or *Plasticine* could be attached to the ends to prevent the reels from sliding off; this can easily be removed to change the arrangement.

CONSTRUCTIONS
1 Use reels for towers and columns. The reels could be left unfixed or glued together to make, for example, a lighthouse, a watchtower, a clocktower, etc.
2 They could be used with other materials in constructions such as lorries (wheels), tanks (undercarriage), caterpillars and dragons' eyes.
3 The form of the reels could be explored by making a unit structure, built one on top of another, placed at different angles to each other. Fix one at a time with an impact adhesive.

PUPPETRY
A reel will make a good neck for a puppet, particularly a rod puppet where the stick can be pushed through the reel before entering the head section. If necessary, the head and neck could be fixed with papier mâché. Or, a piece of fabric could be glued over the join as a dress or a collar. A ball of *Plasticine* or *Newclay* will make a suitable head. See PUPPETS under TWIGS.

PRINTMAKING
Use the end or the side of a cotton reel. The side can be rolled across the paper. See PRINTMAKING under BARK. Try a repeat pattern. Some of the printed shapes could be overlapped and interspersed. Printing with white or a light colour on a black or dark coloured paper can produce interesting results.

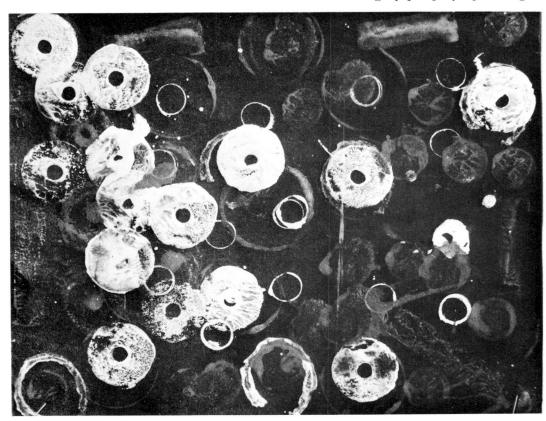

Curtain rings

Collect rings of different materials and sizes.

SORTING
Put the rings into sets according to:
a size
b thickness
c materials (plastic, metal or wood)

27 *Impressions have been made by pressing curtain rings into* Newclay *by a four-year-old. A family of shapes can result and the design be made more interesting by overlapping.* Newclay *can be painted (using powder colour mixed with a PVA binder). Alternatively a print could be taken from the surface. See* Printmaking *under* Bark

PRINTMAKING
Press a ring(s) into a small *Plasticine* block or glue to a wooden block or strong cardboard box. Use a sponge pad for printing. See PRINTMAKING under SPONGE.

IMPRESSIONS
Press a ring(s) into a block of *Plasticine, Newclay,* or *Play dough.* Make a design by interspersing and overlapping. Remove the ring(s), leaving the impressions in the surface of the block. See IMPRESSIONS under BOTTLE CAPS.
Print from the impressions. See PRINTMAKING under BARK.

RELIEF
Make a design using different sized rings. Children enjoy playing with rings, making different arrangements. When they find patterns they wish to keep, the rings can be stuck to the board or card base. Rings can be overlapped and built up into several layers. An impact adhesive or PVA binder should be used.

Dowelling rod

Collect off-cuts.

PUPPETRY
Use a short length of dowelling as a rod for a puppet. See PUPPETS under COTTON REELS and TWIGS.

MASKS
See PAPIER MÂCHÉ MASKS under BALLOONS.
2   Card masks, see MASKS 2 under CANE.

RELIEF
Arrange dowelling rods of the same or different lengths and thicknesses into a design, and glue to a base board.

CONSTRUCTIONS
1   Make a free standing form, fixing the lengths of dowelling rod together one at a time with an impact adhesive. Structures could be made without an adhesive by overlapping and interweaving.
2   Use a board base to make a construction with the dowelling rods standing upright, glued at the bottom end to the base. Try to arrange them in a pattern or in some order.

COLOUR BOX
See COLOUR BOX under CARDBOARD BOXES.

DRAWING STICK
See DRAWING under SPONGE.

29   *A drawing stick is being used here by a three-year-old. The stick consists of a piece of sponge inserted into a slot at the end of a short length of dowelling rod*

28   *Stick 'faces' made by a six-year-old from a paper plate. They were painted with powder colour, decorated with wood shavings and a dowelling rod inserted into an Aloplast base*

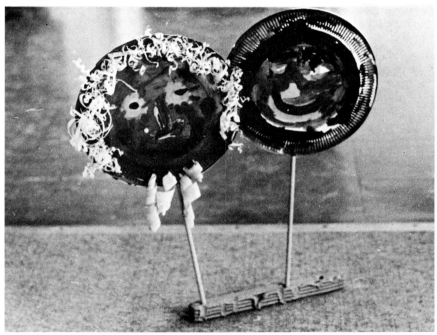

30 *An egg box mask made by a three-year-old child. It was coloured with felt tip pens*

STORAGE
Use as trays for collected items. They can be lined with tissue paper or cotton wool for fragile substances.

PAINTING/DRAWING
The inside and the outside of the boxes can be painted with water based colour *eg* powder colour, temperapaste, and poster colour. Inks can be used. Felt tip pens and colouring sticks (non-toxic) are good for drawing, as well as pastels and wax crayons.

PRINTMAKING
Use the base of an individual segment, held in the hand, for printing on paper. A repeat pattern or a pictorial image could be tried. Marks printed in this way could be combined with those from other materials. A sponge pad is the most suitable when printing in this way. See PRINTMAKING under SPONGE.

RELIEF
Arrange whole boxes, half boxes, and individual segments into a design on a board or card base. They could be placed close to each other to cover the whole of the base, or spaces could be left between them. The spaces could be decorated with paint or collage. Layers can be built one on top of another.

CONSTRUCTIONS

1 Whole or half boxes can be stuck together one at a time. An impact adhesive is the most effective, since it is important that the boxes are firmly glued before adding more to the structure.

2 Large structures can be made by sticking egg boxes to the surface of cardboard boxes. An exciting totem pole can be made in this way. Individual segments stuck onto a cardboard box robot make convincing eyes and nose. The egg boxes can be decorated by painting, drawing, printing, and collage.

3 Use individual segments to make a unit or molecular structure. Link them together with matchsticks or cocktail sticks passing through small holes in the sides of the segments. Glue on the inside.

4 Make a construction joining half boxes together by slotting and intersecting (cut slits in the sides).

31 *A caterpillar made by a three-year-old from individual egg box segments linked with wool. They have been painted with powder colour.*

Matchsticks form the antennae and fabric strips have been glued to the side for the eyes, nose and mouth

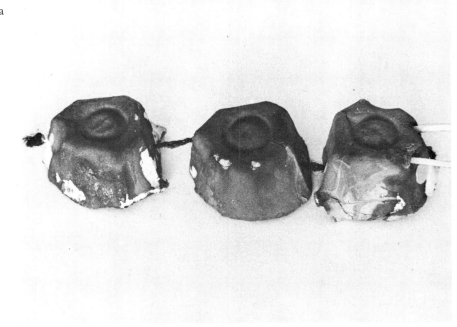

Egg shells

RELIEF

Use half shells together or in combination with smaller pieces. Try placing the half shells at different angles. They could be either way up, or turned on their sides. Glue the touching surfaces. A richer overall texture will be achieved by using smaller pieces of egg shell. Some half shells could then be used for contrast, projecting higher from the base board or card (side of a waste cardboard box). The visible surfaces can be coloured with water based paint or inks (use a soft haired brush *eg* squirrel hair) or by drawing with a felt tip pen (non-toxic). Make sure the glue has dried and surfaces are firmly stuck before colouring.

32 *Drawing the face of a snake on an egg shell using a felt tip pen. The shells have been glued together with* Caretex

Elastic bands

TIE AND DYE

Mask areas of the fabric from the dye by tying round with elastic bands. The tighter the bands are the more effective they will be. They are particularly useful for young children who may not be able to tie knots in string tightly enough. See TIE AND DYE under FABRIC SCRAPS and STONES and the bibliography for further information.

MODELLING

See MODELLING under NEWSPAPERS.

SCREEN PRINTING

Use elastic bands to hold silk, nylon or cotton organdie taut across a kilner jar lid or a section of a cardboard tube. See PRINTMAKING 2 under NYLON STOCKINGS and SILK SCREEN PRINTING under WAX.

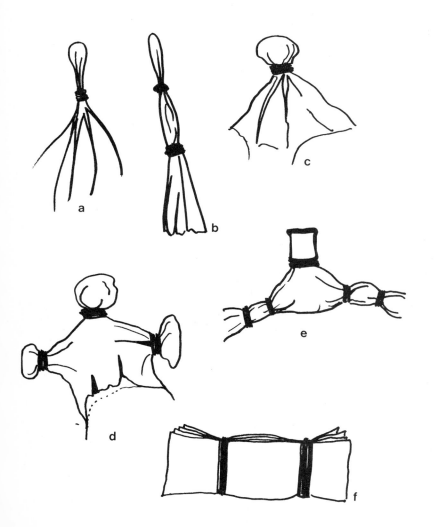

33   *Suggestions for the use of elastic bands or (string) in tie and dye*

a   *Fabric gathered at centre and one tie made with elastic bands*
b   *Two areas are masked from the dye*
c   *Pebble tied into cloth and tightly bound in place with elastic bands*
d   *More than one pebble used*
e   *Central area of fabric gathered and tied round a cotton reel. Four other places are tied*
f   *Cloth is pleated before tying tightly with elastic bands*

Fabric scraps

Keep a large box handy for scraps and discarded clean clothes. As the collection grows and the variety of fabrics increases, it is a good idea to sort into different boxes.

34 *Fabric scraps have been used by a six-year-old to make this picture of children playing in the garden. The scraps were glued to a paper base with* Caretex

SORTING
See SORTING under BEADS. Sort into boxes according to
a colour
b texture
c material (*eg* nylon, cotton, hessian and velvet)

DISCUSSION
Where do the materials come from? What is the difference in origin between cotton, wool, and nylon? Which do you most/least like to touch?

Descriptions by the children of what the fabric surfaces feel like.

APPLIQUÉ
Use a thick, strong material such as felt, hessian or velvet as a base on which to stick smaller pieces of fabric. *Caretex* is a good adhesive for absorbent materials. Older and more manually capable children could sew pieces to the base cloth.

COLLAGE
1   Use fabric scraps in combination with papers and card in a collage on a paper or card base.
2   Surfaces of constructions, balloon faces, masks, and three-dimensional work in general can be decorated with fabric scraps.

RUBBINGS
See RUBBINGS under WAX.
1   Use newsprint and a coloured wax crayon to take rubbings from appliqué and collage work.
2   Individual pieces of fabric may offer interesting textures to include in a collection of rubbings; these rubbings may be used in future collage work or in a frottage picture.

PUPPETRY
Fabric scraps will be in constant demand if undertaking puppetry work on a large scale. Among the simplest are glove puppets; also puppets made from stockings or tights stuffed with old newspapers or rags, tied or sewn up, then decorated with fabric scraps. Further information can be found in the bibliography at the end of this book. Useful starter ideas can be found in *Introducing Puppetry* by Peter Fraser, published by B T Batsford Limited.

PAINTING/DRAWING STICK. See DRAWING under SPONGE.

PAINTING
Use fabric as a paint applicator – dabbing, sweeping, rotating, wiping with the fabric loaded with water based paint onto a paper or card surface. Many exciting marks can be made which can be used in picture-making, patterns and designs.

PRINTMAKING
1   Print with individual pieces of fabric with hard or strong textures, painted or rolled up with water based colour. See PRINTMAKING under CARDBOARD BOXES. A sheet of newsprint is placed on top of the fabric, and pressed with the hand to take a print.
2   The fabric could be glued to a block for hand held printing. Try making a pattern. See PRINTMAKING under BUTTONS.
3   Fabric as a printing surface. If printing an overall pattern the fabric should be laid flat on a firm surface, smoothed out and held in place with *Sellotape* or drawing pins. (If the fabric is creased iron before using.) Otherwise a satisfactory print can be obtained without fixing the fabric into place. There are many different ways of printing on a fabric surface, most of which are best approached in later years. (See bibliography for further information.) Some of them can be made simpler for the young child, and equipment can be

improvised instead of using the more complicated and often expensive purpose designed tools. See PRINTMAKING under POTATOES and NYLON STOCKINGS.

TIE AND DYE
Pieces of plain, preferably white, cotton are best: sheeting and old shirts are suitable. Cut or tear into the desired shape and size. Mask or protect parts of the fabric by gathering up, bunching, and tying with string or elastic bands, or by using clips, *eg* paper and bulldog clips and clothes pegs. This will protect parts of the fabric when it is placed in the dye. It is important to read instructions carefully before using any of the different dyes. Cold or hot water dyes should be used under supervision. 'Fast' dyes can be purchased or usually dyes can be made 'fast' by adding salt. See TIE and DYE under STONES and CLOTHES PEGS, and the bibliography for further information.

PAPIER MÂCHÉ
Strips of fabric can give strength to papier mâché work. See PAPIER MÂCHÉ under BALLOONS.

LIGHT EXPERIMENTS
See LIGHT EXPERIMENTS under CARDBOARD BOXES.

Feathers

Encourage children to collect feathers they find on the ground.

SORTING
This might be according to:

a   size
b   shape
c   colour
d   pattern
e   texture

See SORTING under BEADS.

35   *The idea for this piece of work came after a walk in the country. An eight-year-old boy collected the feathers, arranged and glued them to a card base using a PVA binder. The eye is a bead*

PRINTMAKING
1   Place a well shaped feather onto a sheet of paper. Roll water based printing ink over the feather. See PRINTING under CARDBOARD BOXES. Either print without moving the feather, or place it on a clean piece of paper. Gently press with the hand another sheet of paper (newsprint) over it to take a print.

2   Squeeze a little water based printing ink onto a tile – formica, lino, a wooden board (coated with a seal or varnish to make it non-absorbent). Roll out the ink so that it covers the tile with an even consistency. Place the feather(s) on top of the ink. Take a print by covering the feather and tile with a sheet of paper and pressing with a clean roller or the back of a dessert spoon. The colour of the paper will be preserved where the feathers have masked it from the ink.

36 (opposite above left)   *Water-based printing ink has been squeezed from its tube onto a tile and Claire, aged three, is rolling it out to an even consistency*
37 (opposite above centre)   *The feather is charged with colour by rolling the ink across its surface*
38 (opposite above right)   *The feather is put onto a clean sheet of paper before the printing paper is placed over it*
39 (opposite below left)   *The paper to receive the print is carefully placed over the feather. Note that the roller has been placed safely back onto the tile out of the way*
40 (opposite below centre)   *Pressure over the back of the paper is applied with a clean roller. The paper is held in position with the other hand*
41 (opposite below right)   *A print taken from a feather using this method*

COLLAGE

A variety of feathers will make the work more exciting. Arrange them into a design on a strong piece of paper or card base. Spread a clear glue on the base in appropriate areas and place feathers on to it. Other materials could be used in combination with the feathers.

APPLIQUÉ

Feathers could be used in appliqué work on a fabric base.

DRAWING/PAINTING

1 Draw with the quill, using inks or water based paint. Stiffer quills are best and they can be sharpened.

2 Use the feathery end to paint with. If slightly dry paint is used some delicate and interesting marks can be made. Let the children discover what different kinds of marks they can make. Newsprint or sugar paper will suffice, cartridge paper is best.

Use silver papers, and baking foil.

ENGRAVING

Lay a sheet of foil on a base board. A pad made from sheets of newspaper between the foil and the base helps to cushion the foil while working on it. Draw across the foil using a cocktail stick, the wooden end of a paint brush or some other suitable instrument. Impressions will be left in the foil. They could remain as they are without further working or they could be rolled over with printing ink

or paint and a print taken. See PRINTING under CARDBOARD BOXES. A rubbing might be possible if gentle pressure is applied when using the wax crayon. See RUBBINGS under WAX.

42 *Pie trays cut to the required shapes and milk bottle tops have been stuck to baking foil by a seven-year-old to make this picture of flowers. The baking foil was first folded over a hardboard offcut*

MODELLING

Make models by folding, bending or screwing a sheet of foil into forms which may be free standing or fixed to a base.

PAINTING

Explore the linear qualities in screwed-up and folded pieces of foil. Unfold the foil then float waterbased ink or paint across the surface. The colour will follow the creases in the foil and a linear pattern will result.

COLLAGE

Apply foil to the surface of constructions, using an adhesive. Most adhesives will work with paper foils. The reflective quality of the foil gives an added interest to the appearance of two- and three-dimensional work. See COLLAGE under FABRIC SCRAPS.

PRINTMAKING

Interesting prints can be made on sheets of foil and compared with prints on papers and fabrics.

43   *This swan is made of baking foil folded and twisted into the required form*

Formica

Use offcuts and scraps. Some working surfaces are of formica and children can paint and print directly onto them. They are easily cleaned by wiping with a damp cloth.

FINGERPAINTING
Young children enjoy painting with their fingers. Formica topped tables are ideal to work on directly with powder colour mixed with cold water paste, *Polycell*, or starch. Offcuts of formica serve as good tiles for finger-painting. Tip a little colour onto the formica so that the children can work in it with their fingers. Aprons should be worn. See MONOPRINTING below.

MONOPRINTING
Use formica as an 'ink plate' for printmaking.
1   A finger painting can be recorded by taking a print. Place a sheet of newsprint over the painting. Gentle hand pressure over the back will be sufficient to take a print. Pull up the paper. Hang or pin up to dry.
2   Subtraction monoprint. Paint or roll up a formica tile (offcut) with a water based colour or printing ink.
a   Draw into the colour, scratching lines and areas away with a pencil, matchstick, wooden end of a paint brush or a twig. Then take a print.
b   Place readymade or natural objects, *eg* an old glove, a doily or a leaf, on top of the ink. Cover with a sheet of paper and take a print. Or, cut or tear shapes in paper, card or fabric to mask off areas of colour.
3   Paint or draw a design on a formica tile with a water based colour (mix with a little starch or *Polycell* to retard drying). Take a print with newsprint as in 1 above. Different tools will produce a variety of marks and lines.
4   Paint or roll up a formica tile with water based colour or printing ink. Lay a sheet of clean newsprint on top of it. Draw over the back with a pencil or ball point pen. Carefully pull up the paper and hang or pin up to dry. The pressure when drawing will take up ink from the tile to give a print.

RELIEF
Small pieces of formica can be arranged into a design and glued onto a base board. This should only be attempted when the pieces of formica are ready to hand.

DRAWING OR PAINTING BOARD
Formica will provide a firm, smooth support for drawing or painting paper. A clip (bulldog) could be used.

Fingerpainting on a formica topped table using Finart Finger Paint *which can be purchased ready mixed to use*

Furniture pieces

Collect unwanted oddments and broken wooden furniture. A strong box will be needed for storage.

RUBBINGS
The surfaces of older furniture are often decorated with carved designs. Also the natural grain of the wood will give an interesting rubbing (except where it has been heavily painted or varnished). See RUBBINGS under WAX.

IMAGINATIVE PLAY
1   Pieces of furniture can extend the possibilities of imaginative play. Legs from chairs may become towers, bridges, walls to farmyards or castles, creatures and monsters.
2   Constructions can be made with odd pieces of furniture in building games – fixed with glue or simply rested on each other.
3   Paint and attach materials to the surfaces of furniture parts to give them more meaning as imaginary objects.

PAINTING
Wooden panels and old wooden chair seats make satisfying paint surfaces.

PRINTMAKING
Use grained and carved surfaces. See PRINTMAKING under CARDBOARD BOXES and BUTTONS. The paper to be printed can be placed over the wooden surface, or the wooden piece, laden with colour, pressed onto the paper.

RELIEF
1   Wood panels make very good bases for relief designs.
2   Arrange smaller pieces into a design and glue to a base board.

ASSEMBLAGES
Collect furniture pieces that look well together and arrange them inside a wooden fruit box or a cardboard box. Fix them into position with glue. When the adhesive is dry, the assemblage can be painted. It might be used by children in displays or in games with model figures, and stimulate further imaginative play, as well as being an end product in itself.

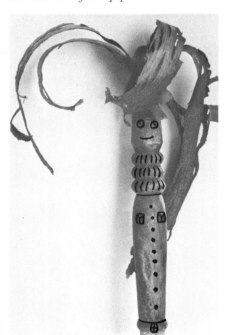

44a   The leg of a chair forms the head and body of this figure, with bark as hair. Contact adhesive is suitable or Evostick Woodworking Adhesive. *Details of face and buttons have been drawn with a felt tip pen*

Greetings cards

DRAWING/PAINTING/PRINTING
Draw, paint or print on clean,
unprinted surfaces.

COLLAGE
Tear or cut the greetings cards into
smaller pieces (training in the use of
scissors and supervision is necessary).
Re-use the pieces in a collage, pasted
onto a paper or card base.

Hardboard

Use offcuts left over from jobs done
in the school and home.

RELIEF
1   Use as a base to designs made with
other materials.
2   Arrange small offcuts into a
design and glue to a base board (a
larger piece of hardboard).

CONSTRUCTIONS
Use small offcuts as sheet material to
make a free standing structure which
could be slotted or clipped together,
rested on top of each other or fixed
together with an impact adhesive.
The structure may be enjoyed for the
colour, texture and shapes of the
hardboard pieces and the form
created. Or it may be used
imaginatively as a shed, barn, etc.

DRAWING BOARD
The smooth surface will make a good
drawing board. Sandpaper the rough
edges. Use a bulldog clip, a clothes
peg or *Sellotape* to hold paper steady
on the board. A piece of hardboard
300 mm by 400 mm (12 in. by 15¾ in.)
is a useful size.

PAINTING
Either side of the hardboard could be
used but it should first be coated with
emulsion paint. The smooth surface is
easier for young children to use.

45   *A hardboard offcut has been used
by a six-year-old as the base for this
bark relief. The smaller pieces of bark
were turned over to make a contrast
with the darker outer surface. The
adhesive was* Marvin Medium

Collect all kinds, clear or coloured, textured or plain. All sizes and shapes will be useful.

46 *The jar inside this lantern holds a night light or candle*

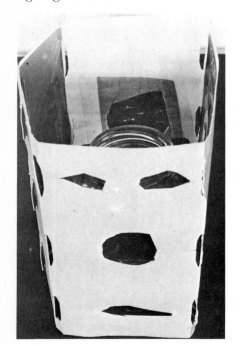

STORAGE

Beads, corks, matchsticks, and many of the other items listed in this book can be usefully stored in jars (an alternative to boxes). Screw tops are important when storing pastes (cold water paste, *Polycell* and glues in jars) to keep them air tight. Dry and liquid paint ready mixed for use is best stored in clear transparent jars so that the colour is easily recognisable.

WATER POTS

With training young children are capable of using glass jars safely. If there is enough space on the working surface, have two or three water pots available to save changing the water too frequently. (Non-breakable pots such as plastic beakers and discarded yoghurt containers are safer. Those with a wider base diameter than top are not toppled over so easily.)

ROLLING PINS

Straight sided jars can be used for rolling out clay, *Newclay*, and *Play dough*. However, a purpose-made rolling pin is much more suitable and safer for young children, and failing this, they can flatten out the clay with hand pressure alone.

LANTERNS

See LANTERNS under CARDBOARD BOXES.

The variety is immense, including leaves of trees, bushes, vegetables, flowers, ferns, reeds and grasses.

DISCUSSION

This can help to develop a critical awareness of the nature of the leaves. Questions about sense responses will further a growing knowledge:
What does it feel like?
Can you describe the shape?
Which leaves do you like best for their colour?
Which is the largest/smallest leaf in the collection?
Which plants do they come from?
Do any of the leaves have a scent?
Describe the scent.

COLLAGE/RELIEF

Using the same or different kinds of leaf, arrange them into a design and glue to a card or board base. The lid of a large box or tin will make a good base and give a framework to the design. The leaves will eventually die but their life will be prolonged by covering, when the glue is dry, with a clear *Transpaseal*. Should this not be possible, try placing a sheet of cellophane or tissue paper over the design. This could be fixed with *Sellotape* on to the back of the base card. Or the tissue paper could be carefully coated with varnish brushed over it. Coloured sheets used in this way will cause interesting changes.

IMPRESSIONS

Use leaves that have well defined shapes and patterns, *eg* horse chestnut or ferns.

Place the leaf on to a *Newclay*, clay, or *Plasticine* tile. Lay a card or thin board on top and gently press. Carefully lift off the leaf; its impression will remain in the tile. With larger tiles several impressions could be made in one tile with the same leaf repeated or with different leaves, side-by-side, interspersed, or overlapped.

RUBBINGS

Making rubbings of
a a single leaf
b a leaf collage (see above)
c a weaving (see below)
See RUBBINGS under WAX.

PRINTMAKING

1 and 2 See PRINTMAKING 1 AND 2 under FEATHERS.
3 Ink a leaf using a roller as in 1. Place a clean piece of newsprint on the table. Pick up the leaf and lay it ink side downward on the paper. Roll a clean roller over it to take a print (a piece of scrap paper over the leaf is advisable). Pick up the leaf, recharge it with ink and repeat the print on a different part of the paper. The prints could be made to overlap. A regular or irregular pattern can be made with several prints of the same leaf.

Different kinds of leaves could be used together in a design.

WEAVING

Use reeds, grasses, stalks of plants (which will be impermanent). Lay lengths of grasses vertically on to a board. Pin or *Sellotape* them down at the ends. These will be the warp. Weave grasses across them. Unpin the ends of the warp and mount the weaving on a card base. Take a sheet of strong paper the same size as the card and cut a window out of it, leaving a frame to be glued over the edges of the weaving. Take a rubbing of the weaving. See RUBBINGS under WAX.

47 *Printmaking with a leaf. Water-based printing ink has been rolled to an even consistency over a tile. A leaf placed on top of the ink masks part of it from the paper on which the print is to be made*

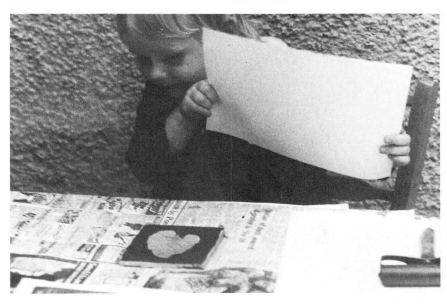

Linoleum

Collect offcuts of lino in the home or school, patterned or plain.

TILES
Use lino for
1 printmaking. See PRINTMAKING under FEATHERS.
2 fingerpainting. See FINGERPAINTING under FORMICA.

WORKING SURFACES
Use especially where a table or bench is of wood and needs protecting (from liquid colour, glue and dye). Cover the table with a flattened sheet of lino which can be pinned, nailed, stuck or clipped into position. Or if it is left unfixed it can be cleared away when the table is needed for other activities.

RELIEF
Use smaller pieces. Arrange them into a design and glue to a board base. The shapes of the pieces will be more distinctive if plain lino is used. Layers can be built one on top of another.

PRINTMAKING
Lino cutters are advisable and are generally too difficult for young children to use. Use a piece of lino as a tile for printing ink. See PRINTMAKING under CARDBOARD BOXES.

RUBBINGS
Take rubbings from linos that have raised or textured surfaces. See RUBBINGS under WAX.

48 *A seven-year-old girl has used different pieces of lino to make this face. The canvas back gives a strong contrast to the darker, smooth lino surface. The lino off-cuts have been stuck to the back of a lino tile with* Marvin Medium

Liquid detergent

SGRAFFITO
See SGRAFFITO under WAX. A little detergent (one part to five parts of liquid colour) helps the water based colour to cover the wax crayon, which normally repels water.

BUBBLE PRINTS
See BUBBLE PRINTS under STRAWS.

Machine parts

Collect those parts that are easily accommodated in the storage space available. Larger pipes and cog wheels may be too cumbersome to store but if at hand children will use them in imaginative play.

SORTING
See SORTING under BEADS.

DISCUSSION
What kind of material is it: metal, plastic or wood?
What machine do you think this came from?
What was the machine used for?
Who made it?
Where was it made?

RELIEF
Washers, clock and watch parts, nuts and bolts, Meccano, and smaller chains are ideal for this. Arrange and glue them to a board base.

IMPRESSIONS
See IMPRESSIONS under BOTTLE CAPS. Smaller machine parts are suitable and give wide scope for making a variety of impressions.

PRINTMAKING
1   Take a print from an impressed tile or block as made above. For printing from a tile use a roller. See PRINTMAKING under CARDBOARD BOXES; for printing from a hand held block use a sponge pad. See PRINTMAKING under SPONGE.
2   Print from pieces of machinery that have interesting surfaces or shapes. See PRINTMAKING under BARK. Small, flat machine parts can be fixed to a block for printing. See PRINTMAKING under BUTTONS.

DRAWING
Use as templates. Place an interesting and easy-to-handle machine part onto a sheet of paper and carefully draw round it. Repeat the process to make a design. Try:
a   overlapping and interspersing the drawings. Colour the new shapes created by overlapping.
b   using drawing instruments of different thicknesses, colours or qualities, eg pencil, ball-point, and a colouring stick.

CONSTRUCTIONS
1   Different machine parts can be used to make a construction fixed together with a suitable adhesive. Plastic material is lighter and easier to handle.
2   Decorate cardboard waste material with small pieces of machinery.
See CONSTRUCTIONS under EGG BOXES, and COLLAGE under BOTTLE CAPS.

49   *A machine part, drawn round many times in different positions on the paper, was the basis for this design by a seven-year-old boy. The new shapes created by overlapping have been coloured with felt tip pens*

Magazines

COLLAGE

Tear or cut the pages. Paste the shapes onto a paper or card base. Use

a colours that look well together (younger children will not be so concerned with colour.)

b shapes that look the same, that fit or nearly fit

c shapes that overlap

d small shapes to give a mosaic effect

WEAVING

Tear the pages into strips. See WEAVING under PLANTS. Use both black and white and coloured pages.

SORTING

Sort the colour pages and torn or cut shapes into colour groups *eg* reds, blues and greens, separated into their own boxes. This, as well as being an exercise to develop colour awareness and categorising, will be useful in collage work. Seven and eight year olds will be able to do this more easily than younger children.

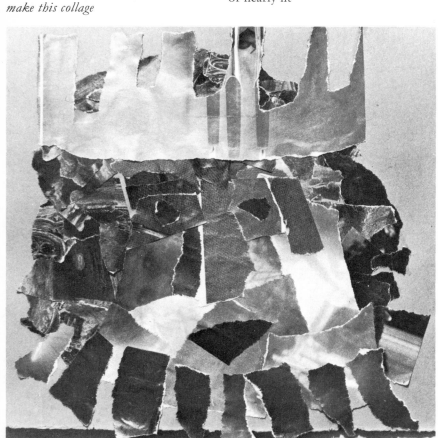

50   *A king. Colour magazine pages, torn and pasted to a card base, have been used by a seven-year-old to make this collage*

Matchboxes

DRAWING

See DRAWING under MACHINE PARTS. Place the matchbox on its ends and sides, in different positions, on a sheet of paper and draw round it.

PRINTMAKING

Use the ends and sides of the tray and the holder. Print by pressing the hand held matchbox on to the paper – newsprint, sugar or cartridge papers, or clean discarded greetings cards and large envelopes. Use a sponge pad. See PRINTMAKING under SPONGE.

COLLAGE
Arrange the opened out and flattened pieces into a design and glue to a strong paper or card base. Make use of the colours and patterns of the labels, and the different textures of the

matchbox, paper wrapper and the striking surface.

RELIEF
Use whole matchboxes (closed and partially open) and the trays and holders separated, to make a design which can be glued to a board base. Place some on end and some on their sides.

CONSTRUCTIONS
1   Use the matchboxes imaginatively, eg as boats, wagons, lorries and cars. Try placing them in different directions to each other.

2   Make unit structures. Glue the matchboxes together one at a time.
3   Make a set of drawers. Glue the boxes side to side and on top of each other. Use brass paper fasteners as handles. The drawers can be used to store small items.

51   Matchbox exteriors glued together by a seven-year-old to make a unit structure

52   Boats constructed from date boxes by seven-year-old children. Matchboxes form part of the structure or when drawn on, become people. The bases of the boxes have been coloured with wax crayons

Matches

Use those without heads or that are dead.

CONSTRUCTIONS
1 Press them into cork, potatoes and other materials being used to make figures, as arms and legs.
2 Use them to join units when making a unit structure. See CONSTRUCTIONS 3 under EGG BOXES.

53 My Cat, *made in matchsticks glued to a paper base by a seven-year-old*

RELIEF
Arrange the matchsticks into a design and glue to a card or board base.

DRAWING/PAINTING
Dip the matchstick into water based ink or paint. Draw or paint on paper.

PRINTMAKING
1 See MONOPRINTING under FORMICA.
2 Use matches in making a 'waste-block'. See PRINTMAKING under CARDBOARD BOXES. It is best not to have too wide a space between.

Metal scrap

Use small, light pieces that are clean and that can easily be glued together or to a base. Possibilities are: keys, hairpins, wire mesh or gauze (from gratings and sieves), cutlery, radio parts and cheese graters.

DISCUSSION
Talk about the qualities of colour, shape, texture, and pattern.
Why are pieces certain shapes?
What is it that makes them a particular colour?
What kind of metal is it: tin, copper, brass, steel or zinc?
What was the object used for?

RELIEF
Choose scraps that fit well together. Try using, for example, only those that are angular, or make a design using only curved pieces. Arrange and glue them to a board base.

CONSTRUCTIONS
Use metal oddments and scraps together or in combination with scraps of other materials, *eg* wood and plastic.
In imaginative play, the different shapes of the scraps may suggest ideas for constructions. Scraps can be stuck together to make a free standing three-dimensional structure.

RUBBINGS

Metal surfaces offer interesting textures for rubbings. Examples are: wire mesh, keys, coins, and insignia on cutlery. See RUBBINGS under WAX.

PRINTMAKING

Ensure that the metal surfaces are clean. Make a printed design with the different shapes and textures. See PRINTMAKING under BARK.

54   *This relief design was made by a six-year-old from clock pieces glued to a cardboard base (a box lid) using a PVA binder*

These should only be used under supervision. It is not advisable to leave nails lying around in the presence of young children. They can, however, serve many useful purposes when used with caution. For use in printmaking or impressions the pointed end of the nail can be made safe and easy to hold by pushing a cork onto it.

SGRAFFITO

Use a nail to scratch away one surface to reveal another below it. See SGRAFFITO under WAX.

PRINTMAKING

See MONOPRINTING 2a under FORMICA. Use a nail with or instead of the items listed there.

IMPRESSIONS

1   See IMPRESSIONS 1 AND 2 under BOTTLE CAPS. Use a nail instead of a bottle cap.
2   Use a nail to decorate clay tiles, pinched pots and creatures modelled in clay, *Newclay*, or *Plasticine*.

54a   *A nail being used to make a pattern round a clay pot*

COLLAGE

1 Use newspapers in combination with other papers.

2 Use them alone to explore the tonal densities of the printed paper. The unprinted areas of the paper will be the lightest (usually white) and the darkest parts of pictures will give the largest areas of black. Different densities of print will offer shades of grey. Tear the newspapers into pieces, arrange them into a design and paste them onto a card or paper base. Black sugar paper is a good base for a tonal collage.

PAINTING

Paint directly on to a newspaper. The print may become part of the painting. Children enjoy the opportunity of working on a large scale where broad marks can be made with the paint. Old newspapers obviate the need to buy larger sheets of cartridge or sugar paper, and give a different surface to work on. (Alternatives are the reverse of wallpaper offcuts and wrapping papers.)

WORKING SURFACE

Use to cover table surfaces for painting, printmaking and modelling.

PAPIER MÂCHÉ

1 Puppets.

a See PAPIER MÂCHÉ under BALLOONS.

b Puppet heads can be made over tennis balls, potatoes and other vegetables. Cover the rounded object with papier mâché (see PAPIER MÂCHÉ under BALLOONS) and allow to dry. Cut round the centre and remove the object from the papier mâché shell. Bring the papier mâché halves together and join with further papier mâché. When dry, a hole could be cut for a neck to be inserted, or the hole could be allowed for in the layers of papier mâché. Decorate when dry, if desired, with paint, colouring sticks, inks or collage. A little PVA binder mixed with paint will give a further strengthening coat to the head.

2 Masks

a See PAPIER MÂCHÉ under BALLOONS.

b Use a suitable modelling material such as *Newclay* to make a relief panel

55 Modelling with newspaper. Elastic bands hold the rolls of newspaper in place. Strips of newspaper and paste can now be applied to give the creature its final form. When dry, it can be painted and decorated

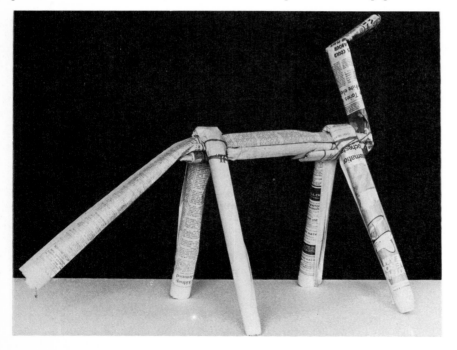

of a face. Allow to dry, then cover with several layers of papier mâché (see PAPIER MÂCHÉ under BALLOONS). When the papier mâché is dry, remove the *Newclay*. The resultant mask can be decorated with collage or colour.

3 Modelling (see below).

MODELLING
Make models with rolled and folded sheets of newspaper, fixed with brown gum strip and papier mâché. Make tight rolls, bend them into the desired shape and tie with string or elastic bands, or stick with brown gum strip. A strong structure can be made which will stand on its own legs. Cover the basic form with papier mâché. See PAPIER MÂCHÉ under BALLOONS. Screwed-up or folded newspaper, or other paper or cardboard waste could be added as projections, to make, for example, a nose or a camel's hump. Several layers of papier mâché will give a smoother finish. When the final layers are dry, decorate with colour and apply different materials as needed, such as fabric scraps for clothing.

WEAVING
Use newspapers torn into strips. See WEAVING under PLANTS. For greater strength use several strips together, perhaps the thickness of one half of a daily newspaper.

Collect those that are discarded and clean, of different shades and denier.

COLLAGE
Cut the nylon into the required shapes, arrange them into a design and stick to a paper or card base. Use a clear adhesive. Try overlapping some of the pieces to create different tonal density.

APPLIQUÉ
Stick or sew the cut shapes on to a fabric base. See APPLIQUÉ under FABRIC SCRAPS. Try
a overlapping the pieces for varying qualities of tone and density.
b contrasting sizes in the design – perhaps use one or two large pieces in a design containing mainly small shapes.

MODELLING
Fill the foot of a stocking with fabric scraps or screwed-up newspapers. The surface can have other materials applied to it by sewing or using a suitable adhesive, *eg Caretex*. The form can be varied by gathering up certain parts of the stocking and tying it round with thread in certain places.

PRINTMAKING
1 Combine nylon with other materials as part of the surface of a 'waste-block' for printing. See

56 *Screen printing with a piece of nylon stocking stretched over one end of a strong cardboard cylinder, held in place with an elastic band. Neil is making a repeat print, masking an area of the paper from the colour with a piece of torn thin card. A foil pie dish is a useful colour palette*

PRINTMAKING under CARDBOARD BOXES.

2   Use in improvised silk screen printing. Stretch a piece of nylon stocking over one end of a kilner jar lid or a cut section of a strong cardboard tube, and hold it fast with an elastic band. Colour can be passed across and through the nylon screen with a piece of card. Powder colour mixed with a little *Polycell* or cold water paste will work well for young children as a substitute for the specially prepared dyes and printing inks. Parts of the paper or card to be printed can be masked from the colour by placing torn or cut paper over it. Shapes can be cut from brown gum strip and stuck onto the screen; these also will act as a mask. Or draw on the screen with wax – a candle or wax crayon. See SILK SCREEN PRINTING under WAX.

INSECT JAR LIDS

Stretch a piece of nylon stocking over a jar and hold in place with an elastic band. This makes a good lid for an insect jar. The experience of collecting and studying different insects is worthwhile in itself. It may well lead into activities such as modelling, making constructions, painting and drawing.

PAINTING

1   Use screwed-up or folded as a paint applicator.
2   Dip into paint and apply as part of a design or picture. Mix a little *Polycell* or cold water paste with the paint. (A relief could be built up in this way. Emulsion paint is ideal for this purpose.)

DYEING

Folded or unfolded, dip into dye or paint. Dip one corner of a folded triangle into a dye. Unfold, allow to dry, and repeat using a different corner and perhaps colour. Try different folds – a square or a pleat folded lengthwise. This can be a way to introduce children to the dyeing process, leading to tie and dye. Note how the colour is absorbed by the paper.

COLLAGE

Use either in their natural state or after dipping into dye, paint or ink. Cut into pieces, arrange into a design and glue to a paper or card base.

CLEANING

They are particularly useful for cleaning table surfaces and brushes.

COLLAGE

Arrange the pipe cleaners and glue to a card or board base. The more they are bent or curved the more difficult they are to fix to the base. Suitable adhesives are *Caretex* or a PVA binder. Use other materials in combination with the pipe cleaners.

PRINTMAKING

They can be used in making a waste-block from which to print. See PRINTMAKING under CARDBOARD BOXES.

CONSTRUCTIONS

They will stay in position when bent or twisted and can easily be manipulated by young children. Use the pipe cleaners as
1   *a*   appendages, *eg* arms, legs, or antennae.
b   links between parts or segments.
c   decoration on constructions and relief work.
2   to make a free standing structure, fixing the pipe cleaners together by linking them around each other and/ or gluing. They are easy to model and can be used to make framework for figures and creatures. Papers or thin fabrics can be pasted across for clothing or wings.

VIBRA-TILES

Fix one end of the pipe cleaners to a firm base. Young children can push them easily into a piece of polystyrene – a tile or waste packaging. To the free end of each cleaner attach a light weight; a piece of card, or pebble, a seed, pasta, or a bead are possibilities. They could be glued or fixed without an adhesive by hooking or twisting the pipe-cleaner round. Use as many pipe cleaners as needed to make the Vibra-tile interesting to watch. Individual stems could be flicked to vibrate, or one already vibrating can be so placed as to set the others into action – the movement will be sustained over a longer period of time. Beads, polystyrene balls or corks are effective weights.

WEAVING

Use the pipe cleaners alone or in combination with other materials. See WEAVING under PLANTS, STRING and SPILLS.

TIE AND DYE

Screw the fabric to be dyed into a ball. Bend a pipe cleaner round it. Twist the ends of the cleaner round each other to hold it firmly in place. More than one pipe cleaner may be necessary to hold the cloth in a bundle. See TIE AND DYE under FABRIC SCRAPS and STONES.

57   Elizabeth makes a flower with pipe cleaners

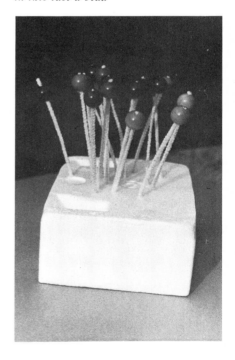

58   A 'vibra-tile'. One end of each pipe cleaner has been inserted in a block of polystyrene packaging material and the other end has a weight attached, in this case a bead

Plants

There is a wide variety of textures, shapes, colours and patterns from which to choose. This can be seen by studying just a few plants: cones, acorns, seaweed, fruit of the horse-chestnut, twigs, rose petals, ferns, reeds and grasses.

DISCUSSION
See DISCUSSION under BONES and LEAVES.

COLLAGE
Using plants which are flat and thin, arrange them into a design and glue to a card or paper base.

RELIEF
Bulkier plant forms, *eg* twigs and seed pods, could be stuck to a base board. They could be imaginatively arranged, or they could be chosen because they have certain qualities in common, *eg* textures (woody or fleshy), shapes or colours.

MOSAIC
See MOSAIC under BEADS. A lid to a biscuit tin would give a strong base and framework; other possibilities are lids to shoe and cake boxes.

Twigs look well when set in plaster of paris. Lay the twigs in the wet plaster. See BASES FOR CONSTRUCTIONS under PLASTIC CONTAINERS.

CONSTRUCTIONS
1   Different plants can be used in many different ways, separately or in combinations. Some examples are:
a   Acorn people or animals. Use matchsticks to join acorns together and as arms and legs. They can easily be pushed into the acorns.
b   Cones with stalks, twigs, or matchsticks inserted as limbs.
Plant parts can be glued together or interlocked.
2   Use nut shells for *a* a relief panel. See under EGG SHELLS.
b   constructions, using broken shells glued together to make a free standing form. Shells could be fixed together with *Plasticine*. An example might be a boat, made with half a walnut shell. Place a small piece of *Plasticine* in the bottom to hold an upright twig or matchstick as a mast. The boat will float. It is important to encourage children to use their own ideas.

PUPPETS
Use twigs as rods for puppets. See PUPPETS under TWIGS.

IMAGINATIVE PLAY
Use woody pieces, *eg* twigs, cones and shells. They have been used as magic wands, rifles, walking sticks, vehicles and machines, built up to make fences and barricades, and used in free play constructions of huts, houses and cots.

A seven-year-old girl has arranged different kinds of plants into a design and glued them to a card base

WEAVING

Use reeds, grasses, stalks of plants (which will be impermanent). Lay lengths of grasses vertically on to a board. Pin or *Sellotape* them down at the ends. These will be the warp. Weave grasses across them. Unpin the ends of the warp and mount the weaving on a card base. Take a sheet of strong paper the same size as the card and cut a window out of it, leaving a frame to be glued over the edges of the weaving. Take a rubbing of the weaving. See RUBBINGS under WAX.

Plastic containers

Collect all kinds including liquid detergent dispensers, cups, yoghurt and margarine pots.

CONSTRUCTIONS
Use whole or parts of containers. They can be glued together with an impact adhesive. The constructions can be decorated with collage and paint. Use emulsion paint or powder colour mixed with a little PVA binder such as *Marvin Medium*.

MOSAICS
Use lids such as those from margarine pots as bases and frames for mosaics. See MOSAICS under BEADS.

SILK SCREEN PRINTING
Use cut sections from taller containers or shallower pots with their bottoms removed. Stretch a fine mesh, *eg* stockinette, across one end and fasten with an elastic band. See PRINTMAKING 2 under NYLON STOCKINGS AND TIGHTS.

BASES FOR CONSTRUCTIONS
Fill a lid or shallow pot with *Plasticine*, *Newclay*, *Play dough* or clay. Insert the legs or the bottom of a free standing structure into the base. For extra strength, the *Plasticine* base can be built up around the legs of the structure. The base can be textured to emphasise the character of the structure.

For a more permanent fixture use plaster of paris. Gradually add the powdered plaster of paris to water in a plastic bowl until it reaches saturation point. At the moment when the plaster begins to solidify, pour the creamy white liquid into a plastic lid. Insert the foot of a construction and hold in place until the plaster has set. The white of the plaster may be preferred to the design and colour of the lid, which can be eased off; or the plaster can be coloured by adding a little powder to the water before mixing.

STORAGE
Plastic bottles such as shampoo and orange squash bottles and liquid detergent dispensers are good for storing ready mixed paint, inks and paste.

PUPPETS
See PUPPETS under PLANTS.

61 *Plastic containers have been used to make a puppet creature with pipe cleaners for legs and antennae. String links the beakers together, knotted in places to keep them spaced apart*

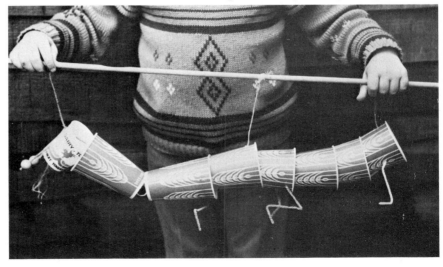

63

There are many different kinds in a variety of forms and a wide range of colours. Plastic bowls, jugs, beakers, cups and plates can all be used. Unwanted articles can be broken up and used in various ways.

IMAGINATIVE PLAY
They will take important roles in children's imaginative play. Have them available in the room, stored perhaps in a cardboard box. They should be clean. Bowls and jugs will be most useful in play with water and sand. Pouring water from a jug into a bowl or a bottle can give pleasure and at the same time the child is developing a concept of capacity and of the properties of the substances he is using.

STORAGE
Use as containers for sand, *Plasticine*, beads and other collected items. Air-tight boxes are useful for keeping *Newclay* fresh. Old bowls and jugs are good for holding mixed paint, inks or paste. Use a plastic bowl for mixing plaster, and *Polycell*.

PALETTES
Plastic saucers or plates can be used as mixing palettes for paint, and to hold colour and a sponge when printmaking. See PRINTMAKING under SPONGE.

MOSAICS
Use the pieces in a design, pressing them into a *Newclay* base. See MOSAICS under BUTTONS.

RELIEF
See RELIEF under BOTTLE CAPS. The shapes of the pieces may help to decide the arrangement for the design.

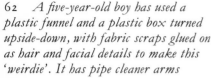

62 *A five-year-old boy has used a plastic funnel and a plastic box turned upside-down, with fabric scraps glued on as hair and facial details to make this 'weirdie'. It has pipe cleaner arms*

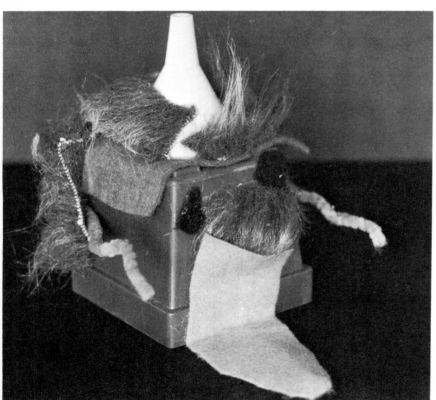

Apply to the surfaces of constructions, for example as eyes, buttons, medals, or to enrich colour and textural qualities.

63 *Daniel is pouring water from a plastic bottle into a plastic jug. In the process of playing he is learning about capacity and relationships of proportions. He is wearing a plastic overall*

64 *A lighthouse made from polystyrene by a group of six-year-olds*

Use waste packaging and oddments such as broken ceiling tiles and insulating panels. A special adhesive will be needed when fixing. Use polystyrene cement, a PVA binder or, with thinner pieces, *Polycell* will serve.

RELIEF
Choose pieces that look well together, arrange them into a design and glue to a base board (hardboard or chipboard, or the lid or side of a cardboard box). The relief could then be painted. The surface can be built up in layers. Pieces can be made to touch side by side or be spaced out.

PAINTING
Use water based inks, dyes or paints (powder colour, tempera paste or emulsion paint work well). Paint directly onto the polystyrene surface.

PRINTMAKING
See PRINTMAKING under BARK
1   Use single pieces exploring the textural quality. Try arranging the prints into a repeat pattern.
2   The surface of the polystyrene can easily be marked or engraved. A finger nail, ball-point-pen-barrel, pencil, twig or matchstick make clear marks and lines. Print the engraved design.

BASE OR MOUNT
Use sheet polystyrene as bases or backgrounds to displays and constructions.

CONSTRUCTIONS
Make a free standing structure, fixing pieces together one at a time. Use of an adhesive may be difficult. *Scotch Construction* adhesive and *Scotch Spra-ment* are recommended. Thinner sheets may already have notches cut or can be cut or torn to make interlocking possible.

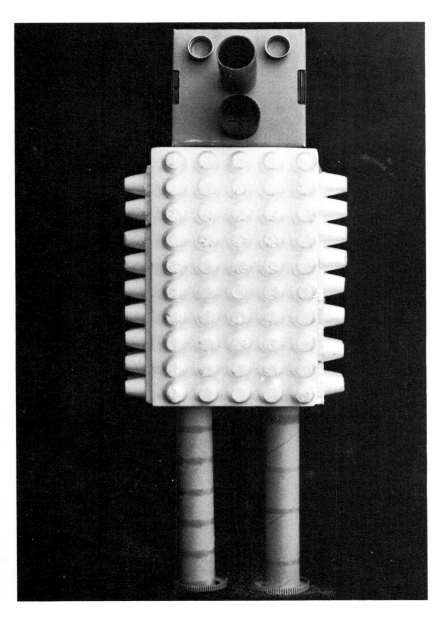

65 Polystyrene tiles glued to a cereal packet form the body of this robot made by a six-year-old. His head is a cardboard box. A toilet roll centre is used for his nose and bottle tops for eyes and mouth. The legs are paper towel roll centres and the feet are jam jar lids. A PVA binder is a suitable adhesive for all these materials

Potatoes (and other vegetables)

See DISCUSSION under
VEGETABLES.

CONSTRUCTIONS
Use a potato as the central body into
which other materials are pressed. A
form or figure can be made by
building on to the potato.
Matchsticks or small twigs will make
satisfactory legs and arms. Feathers,
leaves and cocktail sticks will easily
penetrate the surface skin. Notches
could be cut into the skin for surface
decoration, perhaps as eyes and a
mouth.

PRINTMAKING
See PRINTMAKING under BARK.
Use a section of a potato. Print from
the flat surface. Repeat the print to
make an all over pattern. Parts of the

66   *Potato prints by a three-year-old
girl*

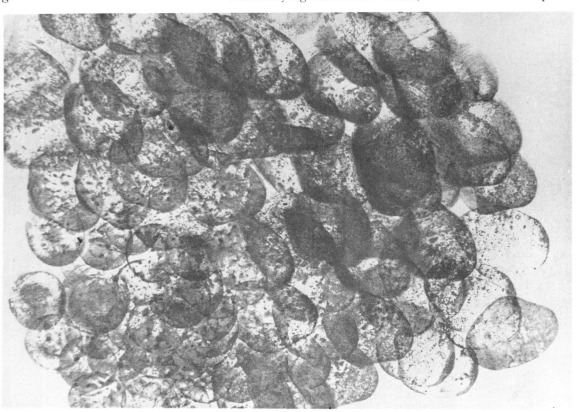

surface of the section could be cut away to leave a motif and printed in a variety of patterns. Each pattern could be in one colour or a combination of colours. The printed potato shapes could be overlapped and interspersed.

PAPIER MÂCHÉ
See PAPIER MÂCHÉ under NEWSPAPERS and BALLOONS.

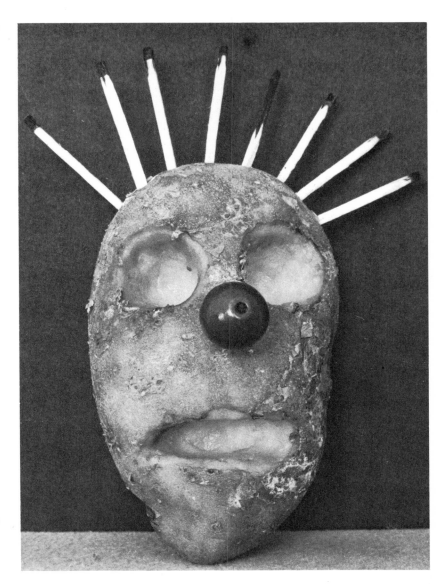

67   *A potato man made by a seven-year-old. The eyes and mouth have been dug out with a spoon. Matchsticks have been inserted to suggest hair. For the nose, a bead has been pushed over the end of a matchstick partially inserted*

Rug oddments

Collect small offcuts of rugs and carpets.

SORTING
Sort according to colour and texture.
See SORTING under BEADS.

DISCUSSION under BEADS and
FABRIC SCRAPS.

RUBBINGS
See RUBBINGS under WAX.

RELIEF
Use shapes that fit and colours that
look well together. Arrange them
into a design and stick to a base of
cardboard or hardboard. Layers could
be built up with pieces overlapping.

PRINTMAKING
1   Use individual pieces chosen for
their textural qualities.
See PRINTMAKING 1 AND 2 under
FABRIC SCRAPS.
2   Use in a 'waste-block' print. See
PRINTMAKING under CARDBOARD
BOXES.

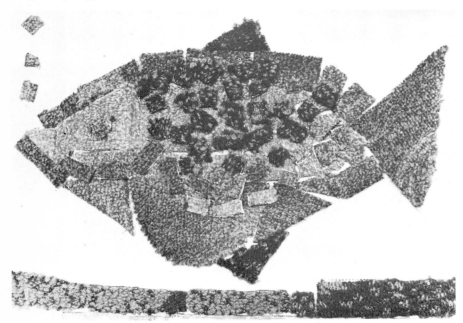

68   *A fish relief panel using small offcuts of rug fabric*

Sand

There are different kinds and colours of sand. To have a variety would make sand play more interesting for the children. See SAND page 10.

PLAY ACTIVITY

1 Filling jugs and bowls, pouring from one receptacle to another, will help develop manual dexterity and rudimentary concepts of capacity and the qualities of substances.

2 *a* Model with the sand.

b Use the sand as a base board on a sheet of PVC for playing out imaginative stories *eg* a beach, a desert, a battleground perhaps incorporating model cars, lorries and figures.

DRAWING

1 Draw with a finger or a stick directly into the sand. This is most effective when at the seaside.

2 Draw with glue on a sheet of paper. Squeeze glue from a tube or paint on with a brush. Before the glue dries sprinkle sand (or a suitable alternative) over it. When the glue has dried gently shake off excess sand. Possible alternatives are glitter, grit and sawdust.

RUBBINGS

Take a rubbing from the sand drawing. See RUBBINGS under WAX.

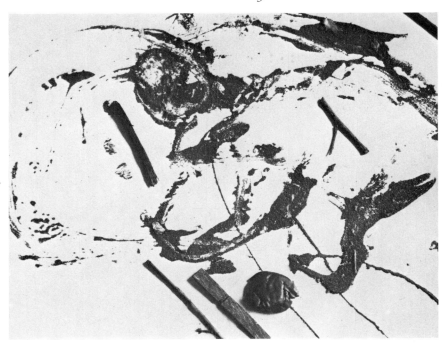

69 *Twigs, a conker and sand sprinkled over wet glue by a four-year-old boy*

Scraps

Scrap material of all different kinds can be used.

SORTING
See SORTING under BEADS, FABRIC SCRAPS and MACHINE PARTS.

COLLAGE APPLIQUÉ
See under FABRIC SCRAPS and BOTTLE CAPS.

MOSAIC
See MOSAIC under BEADS.

RELIEF
Arrange the chosen scraps into a design on a strong base, introducing different textures and thicknesses. A PVA binder will fix most materials.

RUBBINGS
See RUBBINGS under WAX. Collect rubbings from the surfaces of different kinds of scrap.

PRINTMAKING
1 Use single pieces. See PRINTMAKING under BARK, BOTTLE CAPS and BUTTONS.
2 'Waste-blocks'. See PRINTMAKING under CARDBOARD BOXES.

70 *Waste block printing. An eight-year-old boy used matchsticks and a milk bottle top stuck to a wood offcut base to make this waste block. Two prints from the block are seen in the foreground*

CONSTRUCTIONS

All kinds of scrap materials can be
combined in imaginative free standing
structures or in abstract forms. Fix
the pieces together one at a time,
using an impact adhesive. Note that
special adhesives are necessary if
polystyrene is to be used. See
CONSTRUCTIONS under
POLYSTYRENE.

71 Two people on a see-saw *by a
six-year-old boy. A wooden clock case,
turned upside-down, rocks back and
forth when pushed. Toilet roll centres,
foil, fabric scraps, buttons and jam jar
lids have been used to create the figures*

Sea shells

SORTING

Put into categories according to size, colour, shape and pattern. See SORTING under BEADS.

DISCUSSIONS

See DISCUSSIONS under CORK and BONES. Consider the patterns on the shells.
Where were the shells found?
What kind of creature lived in them?
How were they harvested?

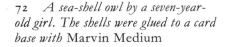

72 *A sea-shell owl by a seven-year-old girl. The shells were glued to a card base with* Marvin Medium

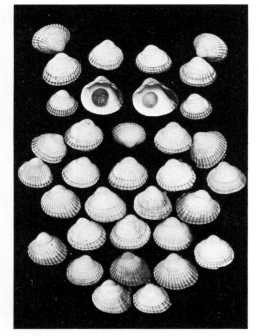

RELIEF

Arrange the shells into a design and glue to a card or board base (hardboard would be suitable). Try placing the shells at different angles to each other.

They could also be glued to the surfaces of constructions.

MOSAIC

Arrange the shells into a design and press into *Newclay*. See MOSAIC under BEADS.

IMPRESSIONS

Use a rolling pin to roll out a piece of *Newclay* or clay into a tile (or use a block of clay). Make impressions in the clay with shells of different sizes and shapes. Clay models and pottery could be decorated in this way. See IMPRESSIONS under BOTTLE CAPS.

PRINTMAKING

1 Take a print of the tile or block of clay impressed as above.
2 Take a print direct from an individual shell. Use either the rim of the shell, which will give a print of the outline shape, or print from the outside surface by rocking the shell across the paper. Use a sponge inking pad. See PRINTMAKING under SPONGE and BARK.

CONSTRUCTIONS

Make a free standing structure by fixing the shells together one at a time, using an impact adhesive.

Seeds (and pasta)

The many different kinds provide a wide choice of colour and size. Fruit pips, such as melon, apple, grapefruit and orange, should be dried before use. They could be used alone or in combination with pasta, lentils, rice and maize.

RELIEF

Spread glue over a card base (a small part at a time) and apply the seeds to it. The design could evolve as the work progresses, or be drawn first on the base before the glue is spread.

MOSAIC

See MOSAIC under BEADS. Pasta is better than seeds for mosaics in *Newclay*. Fruit seeds may swell when moist which is a disadvantage when the surface is painted with a protective coating such as a PVA binder.

JEWELLERY

Seeds dried and pierced can be threaded to make a necklace. This may be an undesirable task for young children since it involves using a needle. Certain pastas (macaroni or spaghetti sticks broken into short lengths) which already have holes can easily be threaded, using cotton, string or twine without a needle.

74   *A necklace and bracelet of melon seeds threaded together with a needle by an eight-year-old boy. The work was done with care and safety*

73 a and b   *Mosaics of pasta pressed into* Newclay *by a seven-year-old. They make attractive paper weights. The* Newclay *was put into the plastic lid of a margarine container and when the pasta and seeds had been arranged, a* coating of PVA binder was painted over the surface. The tub lid may be removed if desired and a piece of felt glued to the back of the paper weight to prevent damage to polished surfaces

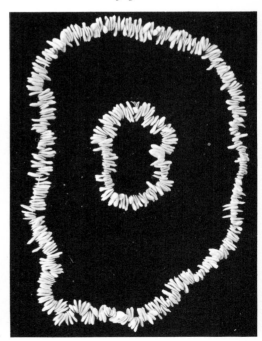

Soap

It is still possible to purchase large blocks of washing (clothes) soap; a tablet of toilet soap would serve well.

CARVING
Carve sculptural forms out of the block. Decorate the surface by scratching into it with various instruments: spatulas, cocktail sticks, 'lollypop' sticks, paper-clips and pencils. The carving could be done with a blunt instrument such as a spoon, a plastic knife or a spatula. Seven and eight year olds could be expected to do this work.

75 *A fish made from a block of soap by a seven-year-old boy. A plastic knife was used for the carving and a cocktail stick and a potato peeler for the surface decoration*

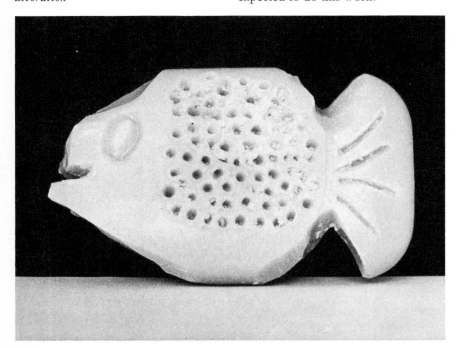

Spills (wood)

Plain wood or coloured spills are available.

WEAVING
Lay spills of the same length vertically on a board. Pin or *Sellotape* them down at the ends. These will be the warp. Weave spills across to make the weft. After five or six spills have been woven across, the pins or *Sellotape* can be removed. The weaving will then be firm enough to hold whilst the weft is completed. Older children may find it possible to weave the spills without fixing beforehand. The woven spills will not need a further support or mount. The completed weaving will make an attractive table mat.

RELIEF
Arrange the spills into a design and glue to a base board. They can be built up into layers, placed flat or on edge, and in different lengths.

CONSTRUCTIONS
Spills can be used together or in combination with other materials. Assemble the spills, gluing them together one at a time. Use an impact adhesive.

They might be used as appendages to other constructions, for example as arms, legs, rifles, pistons and signals; and façades to buildings and vehicles.

Sponge (and foam plastic)

PAINTING

Use a sponge, or a piece of foam plastic folded over and tied with string or an elastic band, as a paint applicator. Either dab the paint onto the paper surface, in which case the texture of the sponge will be evident, or make broad movements, sweeping the colour across the paper.

DRAWING

Cut a slit in one end of a short length of dowelling rod or cane. Insert a piece of sponge or foam plastic. It can be held in place with an elastic band or string if necessary. This will provide a different kind of drawing instrument for use with inks or paint.

PRINTMAKING

Use as an inking pad. Place a piece of sponge or foam plastic in a tray or lid. Soak the pad with ink, ready mixed powder colour or dye. Press a hand-held object onto the pad to load it with colour for printing.

MODELLING

Use to fill out stockings, socks, bags and gloves in making soft toys, *eg* puppets. It can be used alone or with other materials. It is best when cut into small pieces.

76 *Sorting spills into colour groups*

77 *Painting with a piece of sponge*

Stones

Collect stones from different sites to ensure a good choice from the many different colours, sizes, textures and shapes.

SORTING AND DISCUSSION

See under BEADS, FABRIC SCRAPS and LEAVES.

Discuss qualities of shape, size, texture, colour and place of origin. Describe the shape of the stones, jagged, rounded, angular and pointed. What made the stones the shapes they are?

78 *Sea shells glued to a stone make an attractive sculpture with interesting colour and texture variations to explore and discuss. It could be used as a paper weight. The stone was painted with powder colour mixed with a PVA binder before applying the shells*

MOSAIC

Use small stones. Choose colours that look well together. See MOSAICS under BEADS.

RELIEF

Arrange the stones into a design and glue to a base board using a PVA binder. Contrasts will make the work interesting, for example

a large with small.

b clusters placed close together contrasting with others wider spread.

PAINTING

Paint the stones with powder colour mixed with a little PVA binder, emulsion paint or poster colour. Let the shapes and textures of the stones suggest the subject to be painted wherever possible. Ideas chosen by children have included animals, human faces, plants and monsters.

FOUND OBJECTS

Stones could be collected for their natural interest, preserved in their natural state to enjoy for what they are. Discussions with the children will help to decide which stones to keep. It may be they will fire imaginative thinking by the children interpreting them as objects they know *eg* a stone may remind a child of an animal he has seen.

TIE AND DYE

Place a stone in the centre of a square of white cotton sheeting. Gather the cotton round the stone and tie with string or elastic bands. It is important to tie as tightly as possible. Young children may find it easier to use elastic bands. Dip the cotton into a dye. See TIE AND DYE under FABRIC SCRAPS. Whatever dye is being used, read the instructions carefully.

Until young children can tie with string or elastic bands there are many other easier ways of masking areas of the fabric from the dye, *eg* by using clothes pegs, bulldog clips, paper clips and *Sellotape*. See the bibliography at the end of this book for further information on tie and dye; *Tie and Dye as a Present-day Craft* by Anne Maile, Mills and Boon, is a good source of ideas.

Straws

Collect paper or plastic, coloured or plain straws.

'BLOW PAINTING'
Use a straw to blow blobs of ink or paint across a sheet of paper or card. An easy way to transfer a blob of ink to the paper is by pipette action. Take a straw and put one end of it into a pot of ink. Place a finger over the other end and take the straw out of the ink. Hold it over the paper, release the finger from the end, and the ink held by suction will be released. By blowing the ink, a network of lines may be achieved or tendrils stretching out from a central blob. The blown design may be left as it is, or it may be turned into a picture of a familiar object by adding drawn and painted marks and lines.

BUBBLE PRINTS
Put a little liquid detergent in the bottom of a plastic beaker or yoghurt pot. Add ready mixed powder colour or ink to a depth of about 25 mm (1 in.) from the bottom of the beaker. Use a drinking straw to blow bubbles in the liquid mixture. The bottom of the straw should be just below the level of the liquid. Blow until the bubbles rise above the top of the beaker. Carefully place a clean sheet of paper over the bubbles to take a print. Repeat with one or more colours to make an all-over pattern.

79 'Bubble prints'. Yoghurt pots are used to hold the colour mixed with liquid detergent. Claire is placing the paper over the bubbles for the third time to create an overall pattern

DRAWING AND PAINTING

Use a straw as a paint or ink applicator. The end can be sharpened to a point with scissors or a knife for finer drawn lines (an improvised quill pen).

COLLAGE

Arrange the straws into a design and fix to a paper or card base. The straws could be cut to different lengths.

RUBBINGS

Take a rubbing from a collage. See RUBBINGS under WAX.

CONSTRUCTIONS

1   Straws cut to different lengths could be used to make a free standing structure. The straws can be
a   threaded together using string, thread, fuse wire or cotton
b   glued
c   interslotted – squash an end of one straw to fit into the end of another
d   interlocked – bend the straws round each other, slotting and gluing in places for extra strength
e   tied with string, thread or elastic bands.
2   Use straws in combination with other materials to make imaginative constructions, perhaps representing an animal character or a building. Straws make good façades to model buildings and fences.

JEWELLERY

Make a necklace. Cut the straws into short lengths and thread on to cotton, yarn or a thin strong cord. There should be no need to use a needle. The straws could be coloured (preferably before cutting), using inks or powder colour mixed with a little PVA binder. Plastic straws are stronger than paper straws and are usually ready coloured.

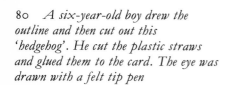

80   *A six-year-old boy drew the outline and then cut out this 'hedgehog'. He cut the plastic straws and glued them to the card. The eye was drawn with a felt tip pen*

String (thread, yarn, twine and cord)

PRINTMAKING

1   Arrange a piece of string into a design and stick to the base of a wood block or to another firm object which can be held in the hand. Take a print. See PRINTMAKING under BARK.
Try winding string round a tube or a bottle and glue or tie it on. Take a print by rolling across the paper.
2   *a*   Dip a length of string (about 700 mm (2 ft 4 in.)) into a water based paint. Arrange it on one half of a sheet of paper. Fold the second half of the paper over the string. Use gentle hand pressure, then unfold. Symmetrical patterns can be made in this way.
b   Place the colour laden string onto the paper so that one end is protruding. After the paper has been folded, place one hand on top to hold it steady, and with the other hand draw the string out, pulling it in different directions. Unfold the paper and allow the print to dry. Protect working surfaces with old newspapers.

COLLAGE/RELIEF

Whether a collage or relief is made depends on the thickness of the string or rope. Thick layers will be nearest to being relief work.
1   Arrange different lengths of string into a design and glue to a board or card base. Cover a small area of the base at a time with glue, then place the string on it. Coloured string will add interest to the design: dip the string into ink, paint or dye and allow it to dry before using by placing it on a sheet of old newspaper.
2   Soak the string in a paste or glue before arranging freely into a design on a card or board base. This can be a messier process than 1 above but it is quicker and perhaps better suited to the more spontaneous working of younger children. Press the string along its length to ensure that it sticks to the base.

WEAVING

1   Wind string round a sheet of strong card; secure it by tying a knot. Weave different thicknesses and kinds of string across it. One long length of string could be woven back and forth, or many separate lengths. Keep the lengths close together. When complete tie back all the loose ends to the nearest length of string crossing them. It will be easier if the edges of the card are serrated or notched.

81   *A string collage of a camel by a six-year-old girl*

Other materials could be used, such as raffia and wool, separately or in combination.

2 Use a wooden frame (an old picture frame would do well) with nails or drawing pins at frequent intervals along the edges. Parents or teachers should do any necessary hammering or supervise nailing or

82 Different kinds of thread have been used to make this 'cobweb' by two seven-year-old girls. The ends of the thread were pulled over slits along the edge of the card and glued to the back

pinning by six to eight year olds. Tie one end of a long piece of string to the end pin on the top edge of the frame and, working from top to bottom, make the warp. One loop round each nail will be sufficient. Weave the weft across with different kinds of string and/or other materials. The side pins and nails may not be necessary if the warp is taut. Tie and trim the ends.

APPLIQUÉ
Glue (or sew) string to a fabric base which should be flat, smoothed out on the working surface. String can add linear, shape, textural and colour qualities to an appliqué in different fabrics. *Caretex* is a good adhesive for absorbent substances.

RUBBINGS
Take a rubbing from the string collage, appliqué or weaving. See RUBBINGS under WAX.

CONSTRUCTIONS
Use string for additional texture to constructions in other materials.

COLOUR BOX
See COLOUR BOX under CARDBOARD BOXES.

JEWELLERY
See JEWELLERY under SEEDS and STRAWS.

SORTING
Sort into sets according to colour, texture, patterns, and/or materials (paper, cellophane, foil).

COLLAGE
Alone or with other materials, arrange the wrappers into a design and glue to a paper or card base. Try overlapping some of them, especially the transparent type.

TRANSPARENCIES
Use the coloured transparent wrappers to make slides. If they are to be used with a slide projector the slides should be 50 mm (2 in.) square. They can be larger for use with an overhead projector.
1 Cut two pieces of card the same size about 250 mm by 197.5 mm (10 in. by 8 in.). Cut a window in the same position out of each. Or use old slide frames. Stick the sweet wrappers together, using a transparent adhesive, so that they make a sheet of wrappers. Lay them between the sheets of cards. Glue round the edges of the card. The sheet of transparent sweet wrappers is now seen through the card window and can be projected onto a screen using an overhead projector or slide projector.
2 Cut two pieces of *Sellotape* (50 mm (2 in.) wide for slides) or clear *Transpaseal* (for larger transparencies) and one sheet of card (manilla card is

Thread

Tiles

suitable) equal in size to the pieces of *Sellotape* or *Transpaseal*. Cut a window in the card leaving a frame of at least 20 mm (¾ in.) wide. Lay one sheet of *Sellotape* or *Transpaseal* flat on the table with the sticky side facing upward. Place the card frame on top of it. Arrange and place the wrappers onto the sticky surface inside the card frame. When the design is complete place the second sheet of *Sellotape* or *Transpaseal* over it. The transparency is now complete and can be projected.

   Other transparent materials could be used in combination with the wrappers, such as tissue paper, nylon, cellophane, leaves, petals and slithers of orange, apple and cucumber skin.

3   Stained glass windows. The larger transparencies described in 1 above could be fixed to a window (with *Sellotape* or *Cow Gum*). This can be done using only one sheet of card, gluing or *Sellotaping* the sheet of wrappers to the card window frame. This would obviate the use of *Transpaseal* which tends to be expensive for class or group use. Again different materials could be used with the wrappers. Tissue paper works well when held against a window.

Various kinds can be collected, including cotton, nylon and elasticated thread. See also under STRING.

MOBILES
Use to suspend card, paper and other materials in a mobile. Thin, strong thread is ideal for this and elasticated thread will give an additional up-and-down motion.

These can include offcuts of formica, plastic, lino, wood and ceramic tiles.

SORTING
Possible categories for sorting are
a   natural materials, *eg* cork and wood.
b   man-made, *eg* plastic and nylon. See SORTING under BEADS and FABRIC SCRAPS.

83   *Sweet-wrappers used to make a transparency to be projected by an overhead projector*

FINGER PAINTING
See FINGER PAINTING under
FORMICA.

84　*A relief panel made from
polystyrene tiles. They have been used
without cutting and were glued to a
board base and to each other with a
PVA binder*

PRINTMAKING
1　Use a tile as an inking plate. See
PRINTMAKING under CARDBOARD
BOXES and FEATHERS.
2　Monoprinting. See FINGER
PAINTING and MONOPRINTING
under FORMICA.

PALETTES
Use the tiles as mixing palettes for
paint (only the non-absorbent
surfaces).

RELIEF
Arrange the tiles into a design and
glue to a base board. The different
materials could be used separately or
in combinations. Polystyrene and
cork tiles will give textural contrasts
to ceramic and plastic surfaces. For
polystyrene suitable adhesives are
*Scotch Spra-ment, Scotch Construction
Adhesive*, Polystyrene cement, or a
PVA binder – a good general purpose
adhesive.

CONSTRUCTIONS
Use the tiles as sheet material.
1　Construct imaginary objects such
as a house, car, fence, garage or shop.
2　Make structures which involve
exploring the materials being used.
Abstract forms can be enjoyed for
themselves and may be used in future
imaginative play.
　　Fix the constructions together by
slotting, interlocking, gluing,
clipping, pinning or nailing.

Collect tins (and their lids) of different
sizes. Make sure they are cleaned well
before use. Labels can easily be
removed by soaking in water. Do not
use tins that have been opened with a
tin opener.

SORTING
See SORTING 2 AND 3 under
BEADS. There is a wide variety of
shapes and sizes available including
mustard, biscuit, powdered milk and
powder paint tins.

DISCUSSION
Consider the reflective surfaces of
certain tins with the children. Effects
of distortion caused by the curved
surfaces might be discussed. Concepts
of capacity could be developed by
considering how much the different
tins can hold. Do you think the short,
fat tin will hold more or less than the
tall, thin one?

MIXING PALETTES
Use the lids as mixing palettes. Either
use individually or make a mixing
palette with as many receptacles as
required (six or eight are good
numbers) by sticking them to a wood
offcut base with a PVA binder.

STORAGE
Use the tins for storing collected
items. Label them with clear, bold
lettering.

RELIEF

1 Arrange the tins with or without lids into a design and glue to a strong base, *eg* a wood offcut (try plywood or chipboard; hardboard may be too pliant). Place the tins at different angles to each other. Explore the various heights and widths of the tins in the design. The base board and the

85 *This tower was built with tins by a seven-year-old boy. It was not fixed with adhesive. The tower was used for different purposes in play activities*

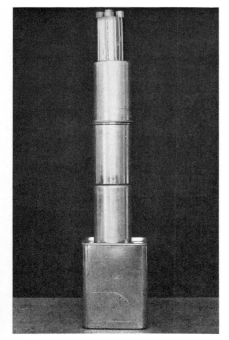

tins can be painted. Keep a check on the weight of the final panel.

2 Use the lids on their own. Arrange them into a design and glue to a base board (hardboard may suffice here).

PRINTMAKING

1 Smaller tins and lids, especially those that have interesting surfaces, can be used to make 'hand-held' prints. A sponge (foam plastic) pad is best. See PRINTMAKING under SPONGE.

2 Wind string, sometimes crossing itself, round a cylindrical tin. Apply colour to the string (brush it on or use a sponge (foam plastic) pad). See PRINTMAKING under SPONGE. Take a print by rolling the tin across a sheet of paper.

CONSTRUCTIONS

1 Glue tins and/or lids together one at a time to make a free standing structure. Use an impact adhesive.

2 Use the tins in combination with other materials: cylindrical, they may become wheels for vehicles or bodies of people or animals; cuboid, they could be used as building bricks, robots, vehicle bodies, or glued together as buildings. The tins can be painted with emulsion or household paint before use.

3 Use a tin as a central body. Stick other materials to its surface with an impact adhesive.

There are many different thicknesses, degrees of transparency, colours and texture. Paper towels, toilet-roll paper, 'tissues' and other purpose-made papers can be used. Art tissue papers, which can usually be purchased from a local art supplier, have a richer and wider range of colours.

COLLAGE

1 Tear or cut shapes and paste them to paper or card. The colour may tend to run when the paper is wet. If so incorporate this into the design. Paste the base paper or card and lay the tissue papers on to it. Use *Polycell* or cold water paste. It may be preferable to put a spot of paste on the corners of each piece of tissue, in which case *Spot-Stick* or *Sellostick* may be useful. Where shapes overlap a greater density will be achieved; and tonal gradations obtained by varying the number of layers. For the best effects in tone, use white or grey paper as the base.

2 Use coloured tissue paper to decorate the surfaces of 'junk constructions'. Examples seen include strips of tissue paper as a horse's mane and tail.

3 Use the tissue in different forms: sheet, strip, crumpled and folded. Paste them to a card or paper base, exploring the different shapes and textures in making the design.

86   *A collage with pieces of tissue paper overlapping each other, creating shapes of darker tones. The paper was cut with scissors by a five-year-old girl and then pasted to a paper base with* Polycell

87   *Torn papers give a different quality compared with the cut pieces in the previous picture. The shapes are more random and the torn edges provide interesting outlines. This is the work of a four-year-old girl who, in pasting the shapes to a paper base with* Polycell, *produced wrinkles and creases which give textural interest*

TRANSPARENCIES
1   Make transparencies for use with a slide projector or an overhead projector. See TRANSPARENCIES under SWEET WRAPPERS.
2   Cut shapes (holes) in a piece of card (some packaging card has shapes already cut out). Tear or cut coloured tissue papers and paste them across the shapes. Place the card against the window.
3   Screw up a piece of coloured tissue paper. Project it onto a screen using an overhead projector. Observe the movement, changing patterns and shapes on the screen as the paper unfolds (cellophane is more effective for colour).

PAPIER MÂCHÉ
See PAPIER MÂCHÉ under BALLOONS.

STORAGE
Use to protect fragile items and to line boxes and trays for storage. See STORAGE under EGG BOXES.

Twigs

DISCUSSION
See DISCUSSION under BARK, BONES and LEAVES.

IMAGINATIVE PLAY
A twig becomes, for example, a walking stick, rifle or a magic wand.

PAINTING
Apply paint with twigs. Try thick paint (of the consistency of toothpaste).

SGRAFFITO
Having painted with a brush, rag, etc, use a twig to scratch away some of the wet surface paint to reveal colours and textures beneath.

DRAWING
Draw with a twig that has been dipped into ink, paint or dye.
Draw into paint while it is still wet, and into mud or sand. Children do this spontaneously on the beach, in the garden, and in sand trays and pits where they are available. The children are gaining control of motor skills whilst enjoying a play activity.

PRINTMAKING
See MONOPRINTING 2a under FORMICA.

RELIEF
Choose twigs that are of suitable lengths for the size of base board to be used. Arrange them into a design and fix to the base. Explore the different colours and thicknesses of the twigs.

CONSTRUCTIONS
1 Twigs can be used to make imaginary structures such as houses, barns, barricades, fences and huts. They can be glued together (with an impact adhesive) or built up by interweaving or overlaying.
2 Use as appendages and surface decorations to constructions in other materials. See CONSTRUCTIONS under POTATOES.

PUPPETS
Use twigs as rods for puppets. See the bibliography on puppetry at the end of this book. One possibility would be to insert a twig through a hole in the bottom of a plastic beaker or yoghurt pot. Make a ball of *Plasticine* or *Newclay* (any spherical object of a suitable material could be used) and push the top of the twig into it. This will be the puppet's head. Other details can be added, buttons, beads, etc., as eyes, nose and mouth pressed into the *Plasticine* or glued to other materials. Fabric scraps could be tied or stuck on as clothing. The twig or rod can be pushed up and down to animate the puppet.

88 *Imaginative play, with a twig used as a pistol*

Vegetables

Use different kinds including potato, carrot, swede, onion, sprout and broad bean.

89   *Printing with a carrot charged with colour by pressing it onto a sponge pad in a saucer of water-based colour. The three-year-old girl repeated the print of the cross-section shape of the carrot many times to produce an overall pattern*

DISCUSSION

See DISCUSSION under BARK, BEADS 2 AND 3, BONES and LEAVES.
Describe the colour, texture and shape.
How was it grown?
How was it harvested: picked, pulled or dug? Were any gardening tools used to collect them?
What does it feel like: hard, soft, dry, wet, rough or smooth?
How does it make you feel when you look at, smell or taste a certain vegetable?
What does it taste like?
Do we use any of the vegetables for anything else besides eating?
Which vegetables are used to make dyes?
Study the growth of the vegetable.

CONSTRUCTIONS

1   Several vegetables could be used in any one construction with twigs or cocktail sticks to link them together. Use firm vegetables.
2   Use one vegetable as a central body to which appendages can be added. See CONSTRUCTIONS under POTATO.
3   Use with papier mâché heads and forms (see PAPIER MÂCHÉ under NEWSPAPERS and BALLOONS).

CARVING

Use a firm vegetable. Possible carving instruments for young children are teaspoons, plastic knives and wooden spatulas. Where older children can be relied upon to use sharper tools safely a potato peeler, apple corer, craft knife or penknife are more effective.

PRINTMAKING

Most vegetables can be used to make prints. Some will last longer than others during the printing process. Parsnip, turnip, carrot and swede are ideal. See PRINTMAKING under POTATO. Use the hand-held method of printing, and a sponge pad. See PRINTMAKING under SPONGE.

Wallpaper (offcuts)

DISCUSSION
Discuss the qualities of colour, texture and pattern with the children.

DRAWING/PAINTING/
PRINTMAKING
This can be done on the reverse of most wallpapers. Nearly all wallpapers are strong enough and suitable for water based colours (washable and vinyl papers also make very good surfaces for oil based colours).

RUBBINGS
Take rubbings from the surfaces of papers that are heavily grained, embossed and rich in texture. See RUBBINGS under WAX.

90 *This mural, made by a family-grouped class of five- to seven-year-olds, consists of collage, drawing and painting on the reverse of lengths of wallpaper placed side by side to give the desired size. Note the use of empty cartons and tins for storage purposes, particularly the use of a biscuit tin to hold jars of ready mixed paint*

Use papers that have interesting
patterns, colours or textures. Cut or
tear shapes of wallpaper, arrange them
into a design and glue or paste to a
paper or card base.

91   *A rubbing made with wax
crayons and kitchen paper by an eight-
year-old. The rubbing was taken from a
collage of a butterfly in wallpaper
oddments.* Polycell *was used to paste
the wallpaper to a paper base*

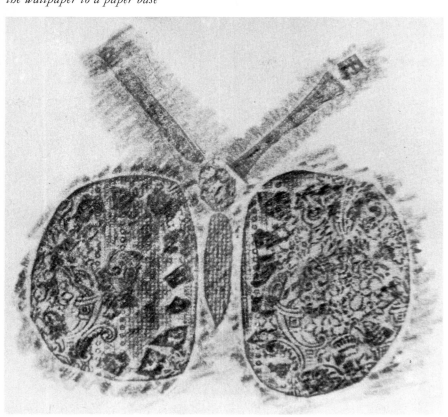

Wax

Use candle wax and wax crayons.

RESIST
Draw with a wax candle (or coloured wax crayon) on a sheet of paper. Paint over the drawing with a contrasting colour of water based paint or ink. *Brusho* colours are recommended. The wax will resist the water based colour and the drawing will be seen through it.

SILK SCREEN PRINTING
Stretch a piece of nylon stocking, stockinette or cotton organdie over a wooden frame (staple or pin the nylon to the frame), kilner jar lid or short section of a strong cardboard tube (tie the nylon round with string or elastic bands). Draw a design onto the screen using a wax crayon or candle wax. Print by placing the screen onto a flat piece of paper or fabric and passing water based colour over the screen. The colour will pass through the unwaxed areas and will print on the paper or fabric. A mixture of three parts powder colour to one part *Polycell* could be used as a substitute for commercial inks or dyes. Use a piece of cardboard or squeegee to move the colour across the screen.

SGRAFFITO
Colour all over a sheet of paper or card (120 mm by 180 mm (4¾ in. by 7 in.) is large enough for young children) with wax crayons. Completely cover the crayoning with black (or strongly contrasting coloured) water based paint or ink mixed with a little liquid detergent (to prevent the wax repelling the water); or cover it with black wax crayon. Scratch a design through the final covering revealing the colour beneath. Possible scratching implements are cocktail sticks, a nail, matchstick, plastic knife or a paper clip.

RUBBINGS
Place a strong, thin paper (newsprint) over an interesting texture: bark, woodgrain, coin or embossed wallpaper. Use a wax crayon (in a contrasting colour to that of the paper) to rub gently over the paper. If necessary the paper can be held steady with drawing pins or *Sellotape*. Rubbings from different surfaces can be cut or torn out, arranged into a design and glued to a base paper or card to make a collage of rubbings.

Seven and eight year olds might try a frottage picture, where the design or picture is created on one sheet of paper by taking rubbings from the different surfaces.

DRAWING
1 Use the ends of coloured candles or wax crayons. Suitable papers are cartridge, sugar and newsprint.
2 Cut notches along the length of a coloured candle or wax crayon and draw with the length across the paper.

92 *A three-year-old girl drawing with wax crayons that have had notches cut along them*

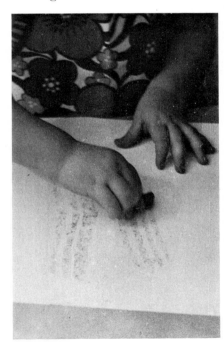

Wood

Use different forms including laths, battening, planks, blocks, shavings and boxes. Use offcuts for most activities.

DISCUSSION
See DISCUSSION under BARK, BONES and LEAVES.

BRICKS
Use offcuts of approximately the same size, cubes if possible. Sand paper the rough edges. Paint the bricks with emulsion paint or powder colour mixed with a little PVA binder. Use the bricks in imaginative structures and games. Try exercises in colour sorting: building towers with bricks of one colour from the collection of different colours.

A child may want to paint a block or piece of wood simply for the enjoyment of painting it.

STORAGE
Stood side by side or one on top of another, wooden boxes can make good shelves for books, games, toys and display areas. Use them to store collected items such as small machine parts, bark, twigs and fabric scraps.

93   A block of wood makes printing with small items easier. Here a button has been glued to the block and is being printed by a three-year-old girl. She applied water-based colour to the button with a brush

CONSTRUCTIONS

1 Place a wooden box so that the open end is facing outward. Inside arrange found objects, bric-a-brac, and items of interest. They can be fixed (use a PVA binder) or unfixed. Several boxes could be used, each containing articles of one kind of material.

2 Make a free standing structure using wood *a* of different sizes and shapes; or *b* of the same size and shape to make a unit structure. In both *a* and *b* glue one piece at a time with an impact adhesive.

PRINTMAKING

See PRINTMAKING under BARK.

1 Print from the grain of the wood.

2 Glue a small item to be printed, *eg* a button or bottle cap, to a block of wood (to be held in the hand). Print using a sponge pad. See PRINTMAKING under SPONGE.

3 Print from impressions made in the surfaces of soft woods such as cedar and birch; balsa wood is also good for this. Impressions can be made with an unsharpened pencil, ball-point pen, an angular stone or a plastic knife.

RUBBINGS

Take rubbings from the grain of different woods. See RUBBINGS under WAX.

RELIEF

1 Use offcuts. Arrange them into a design and glue to a base board: hardboard, plywood, blockboard or chipboard.

2 Wood shavings applied to surfaces for decoration could be used in relief panels and in collage work. They have been used imaginatively by children as hair, clouds, waves and sheep's wool; as well as enjoyed for their own qualities of colour, texture and shape.

PAINTING

A wooden surface, especially one that has already been primed, will be good to paint on. Use emulsion paint or powder colour mixed with a little PVA binder. A wood scrap such as an old chair seat ready primed would be ideal.

94 *Wood offcuts used as blocks for building a construction, glued with PVA binder, then painted with powder colours mixed with PVA binder*

Bibliography

COLLAGE
Creative Collage Ivy Haley
Batsford London, Branford Newton Centre

Pictures with Coloured Paper
Lothar Kampmann, Batsford London
Van Nostrand Reinhold New York

Creating in Collage
N d'Arbeloff and J Yates
Studio Vista London
Watson-Guptill New York

Introducing Seed Collage
Caryl and Gordon Sims, Batsford London
Watson-Guptill New York

Introducing Fabric Collage
Margaret Connor, Batsford London
Watson-Guptill New York

CLAYWORK
Creative Clay Work H Isenstein
Oak Tree Press London

Creative Clay Craft Ernst Röttger
Batsford London
Van Nostrand Reinhold New York

Clay in the Primary School
Warren Farnworth
Batsford London

Working With Clay and Plaster
David Cowley, Batsford London
Watson-Guptill New York

DRAWING AND PAINTING
Introducing Finger Painting
Guy Scott, Batsford London

Introducing Crayon Techniques
Henry Pluckrose, Batsford London
Watson-Guptill New York

Pictures With Inks
Lothar Kampmann, Batsford London
Van Nostrand Reinhold New York

TEXTILES
Fabric Pictures E Alexander
Mills and Boon London

Textile Printing and Dyeing
Nora Proud, Batsford London
Van Nostrand Reinhold New York

Tie and Dye Made Easy Anne Maile
Mills and Boon London
Taplinger New York

Introducing Textile Printing
Nora Proud, Batsford London
Watson-Guptill New York

Introducing Dyeing and Printing
Beryl Ash and Anthony Dyson
Batsford London,
Watson-Guptill New York

MOSAICS
Making Mosaics, Arvois
Oak Tree Press London

Making Mosaics, John Berry
Studio Vista London
Watson-Guptill New York

Mosaics, R Williamson
Crosby Lockwood London

PAPER CRAFT
Creative Paper Craft
Ernst Röttger
Batsford London
Van Nostrand Reinhold New York

Creative Corrugated Paper Craft
R Hartung, Batsford London
Van Nostrand Reinhold New York

Take An Egg Box R Slade
Faber London
Lothrop, Lee and Shepherd New York

Paper Sculpture George Borchard
Batsford London
Watson-Guptill New York

PRINTMAKING
Creative Rubbings Laye Andrew
Batsford London
Watson-Guptill New York

Creative Printmaking
Peter Green, Batsford London
Watson-Guptill New York

Introducing Surface Printing
Peter Green, Batsford London
Watson-Guptill New York

Introducing Screen Printing
Anthony Kinsey, Batsford London
Watson-Guptill New York

PUPPETRY
Introducing Puppetry
Peter Fraser, Batsford London
Watson-Guptill New York

Puppetry In The Primary School
David Currell, Batsford London

Hand and String Puppets
W Lancaster, Dryad Leicester

Puppet Circus Peter Fraser,
Batsford London, Taplinger New York

SCULPTURE
Introducing Constructional Art
Edward Rogers and Thomas Sutcliffe
Batsford London
Watson-Guptill New York

Creating With Plaster D Z Meilach
Allen and Unwin London
Reilly and Lee Chicago

Clay Modelling
Lothar Kampmann, Batsford London

Art From Found Materials
Mary Lou Stribling
Crown Publishers New York

Starting With Sculpture R Dawson
Studio Vista London
Watson-Guptill New York

TOY MAKING
Introducing Soft Toy Making
Delphine Davidson, Batsford London
Praeger New York

Making Felt Toys and Glove Puppets
Batsford London, Branford Newton Centre

MISCELLANEOUS AND
GENERAL CRAFTS
Making Mobiles
Anne and Christopher Moorey
Studio Vista London
Watson-Guptill New York

Introducing Jewelry Making
John Crawford, Batsford London

Introducing Beads
Mary Seyd, Batsford London

Children's Costumes in Paper and Card
Suzy Ives, Batsford London

Making Masks
Barbara Snook, Batsford London

Ideas for Art Teachers Peter H Gooch
Batsford London
Van Nostrand Reinhold New York

Creative Crafts For Today
John Portchmouth
Studio Vista London
Viking Press New York

Pre-School Activities
Dorothy and John M Pickering
Batsford London

THEORY:ART
Visual Education in The Primary School
John M Pickering
Batsford London
Watson-Guptill New York

Pre School and Infant Art
Kenneth Jameson
Studio Vista London
Viking Press New York

Art With Children Daphne Plaskow
Studio Vista London
Watson-Guptill New York

Creative and Mental Growth
Viktor Lowenfield
Collier-Macmillan London and New York

THEORY: GENERAL
Playing, Learning and Living
Vera Roberts, A & C Black London

Play With a Purpose for Under Sevens
E M Matteson, Penguin Harmondsworth

Senses and Sensitivity
Alice Yardley, Evans London
Citation Press New York

Learning through Creative Work
(The Under Eights in School)
Beatrice F Mann
National Froebel Foundation
5th revised edition 1971

Family Grouping in the Primary School
Lorna Ridgway and Irene Lawton
Ward Lock Educational London
Agathon Press New York

Activity Methods for Children Under Eight
edited by Constance Sturmey
Evans London

Modern Practice in The Infant School
Lesley Webb, Blackwell Oxford

Display in The Classroom
Ruth Phelps, Blackwell Oxford

Suppliers

Fred Aldous Ltd
The Handicrafts Centre
37 Lever Street
Manchester M60 1UX

E J Arnold and Son Ltd
(School Suppliers)
10 Butterley Street
Leeds LS10 1AX

Cosmic Crayon Company
Ampthill Road
Bedford

Crafts Unlimited
21 Macklin Street
London WC2

Galt Early Stages
James Galt and Co Ltd
Brookfield Road
Cheadle
Cheshire SK8 2PN

Margros Ltd
Monument House
Monument Way West
Woking
Surrey

Newclay Products Ltd
Overton House
Sunneyfield Road
Chislehurst
Kent

Nottingham Handcraft Co
(School Suppliers)
Melton Road
West Bridgford
Nottingham NG2 6HD

Reeves and Sons Ltd
Lincoln Road
Enfield
Middlesex

George Rowney and Co
10–11 Percy Street
London W1

Winsor and Newton Ltd
Wealdstone
Harrow
Middlesex
and 51 Rathbone Place
London W1